Wolfgang Tutschke

Solution of Initial Value Problems in Classes of Generalized Analytic Functions

Springer-Verlag Berlin Heidelberg GmbH

Prof. Dr. Wolfgang Tutschke
Sektion Mathematik
Martin-Luther-Universität, Halle-Wittenberg
4020 Halle, GDR

ISBN 978-3-540-50216-6 ISBN 978-3-662-09943-8 (eBook)
DOI 10.1007/978-3-662-09943-8

Library of Congress Cataloging-in-Publication Data
Tutschke, Wolfgang, 1934 – Solution of initial value problems in classes of generalized analytic functions/Wolfgang Tutschke.
p. cm. Bibliography: p.
ISBN 978-3-540-50216-6 (U.S.)
1. Initial value problems. 2. Analytic functions. I. Title.
QA378.T87 1989 515.3′5–dc 19 88-24882

Originally published by Springer-Verlag Berlin Heidelberg New York in 1989
Licence edition for Springer-Verlag Berlin Heidelberg 1989

2141/3140-543210

PREFACE

From the very beginning the development of Complex Analysis was closely connected with the theory of partial differential equations. One of the highlights of these interactions was the creation of the theory of generalized analytic functions because it linked complex-analytic and functional-analytic ideas.

Whereas the theory of generalized analytic functions is mainly aimed at solving boundary value problems, initial value problems can also be solved by using complex methods. The classical Cauchy-Kovalevskaya theorem is the first fundamental result in this direction. The functional-analytic approach to that theorem, started by M. Nagumo's ideas, led to an abstract version of the Cauchy-Kovalevskaya theorem.

The present book combines both the above-mentioned trends in Complex Analysis. It contains results partly obtained in the research group "Partial complex differential equations" of the Mathematical Department at Halle University. This group tries to contribute to the further development of the theory of generalized analytic functions.

Many people have supported me in the preparation of this book. I would like to thank all of them. For including this book in the series "Teubner-Texte zur Mathematik", I am grateful to the Teubner publishers and the editors of that series. For constructive co-operation I thank the staff of the publishing house, especially Mrs Dr. Müller and Mrs Roth. I also appreciate Mrs Ziegler's work on the copy-ready manuscript.

I am grateful to the publishing house "Springer-Verlag", especially to Dr. Heinze for his interest in the manuscript. I am very glad that this publishing house participated in the edition of the present book.

I was also supported by many colleagues from the Halle University Mathematics Department. I would like to thank my colleagues, especially Dr. Buchsteiner.

Halle, June 1988 Wolfgang Tutschke

CONTENTS

0. INTRODUCTION

The purpose of the present book is to solve initial value problems in classes of generalized analytic functions as well as to explain the functional-analytic background material in detail. From the point of view of the theory of partial differential equations the book is intended to generalize the classical Cauchy-Kovalevskaya theorem, whereas the functional-analytic background connected with the method of successive approximations and the contraction-mapping principle leads to the concept of so-called scales of Banach spaces:

1. The method of successive approximations allows to solve the initial value problem

$$\frac{du}{dt} = f(t,u), \tag{0.1}$$

$$u(0) = u_0, \tag{0.2}$$

where $u = u(t)$ ist real or vector-valued. It is well-known that this method is also applicable if the function u belongs to a Banach space. A completely new situation arises if the right-hand side $f(t,u)$ of the differential equation (0.1) depends on a certain derivative Du of the sought function, i. e., the differential equation (0.1) is replaced by the more general differential equation

$$\frac{du}{dt} = f(t,u,Du). \tag{0.3}$$

There are differential equations of type (0.3) with smooth right-hand sides not possessing any solution to say nothing about the solvability of the initial value problem (0.3), (0.2). Assume, for instance, that the unknown function denoted by w is complex-valued and depends not only on the real variable t that can be interpreted as time but also on spacelike variables x and y. Then the differential equation (0.3) can be rewritten as

$$\frac{\partial w}{\partial t} = f(t, w, \frac{\partial w}{\partial x}, \frac{\partial w}{\partial y}). \tag{0.4}$$

H. Lewy [35] proved that there exist infinitely differentiable right-hand sides f so that (0.4) does not possess any solution.

The H. Lewy example was the starting point of a lot of papers answering the question whether a partial differential equation possesses at least one solution or none (cf. section 9.3.1.). This question, however, will not be in the limelight. We are going to derive sufficient conditions under which even the initial value problem is solvable.

2. On the other hand the classical Cauchy-Kovalevskaya theorem shows that the initial value problem

$$w(0,x,y) = w_0(x,y)$$

to the differential equation (0.4) is always solvable if the right-hand side of (0.4) and the initial function $w_0 = w_0(x,y)$ are holomorphic functions in all their variables. The classical proof of this theorem is based on the power-series representation of holomorphic functions. The classical Cauchy-Kovalevskaya theorem turns out to be a special case of the different kinds of generalizations.

3. Scales of Banach spaces are certain families of Banach spaces depending on a real parameter. Such scales allow the initial value problem (0.2) to be solved by the method of successive approximations also in the case of differential equations of type (0.3). In this way one gets an abstract version of the Cauchy-Kovalevskaya theorem (cf. T. Yamanaka [80], L. V. Ovsyannikov [50], L. Nirenberg [45, 46], F. Treves [61], T. Nishida [47]). The application of this abstract version of the Cauchy-Kovalevskaya theorem to scales of Banach spaces of holomorphic functions yields immediately the classical Cauchy-Kovalevskaya theorem mentioned above.

4. One of the goals of the present book is to give a detailed introduction into the use of scales of Banach spaces for solving initial value problems. The main goal of the book is, however, to apply this method to scales of Banach spaces of generalized analytic functions. Generalized analytic functions are solutions to elliptic systems of differential equations in the plane or in higher-dimensional Euclidian spaces generalizing the well-known Cauchy-Riemann system. That way we shall be able to solve initial value problems with generalized analytic functions as initial functions.

5. The classical Cauchy-Kovalevskaya theorem is based on the fact that the derivative of a holomorphic function is holomorphic, again. Generalizing this property of holomorphic functions to the case of generalized analytic functions, one gets the concept of associated differential operators.

6. Using the concept of the so-called dual scale, the classical Holmgren uniqueness theorem can also be deduced from the abstract Cauchy-Kovalevskaya theorem. The latter is, consequently, advantegeous for the classical theory, too.

7. The functional-analytic approach to the Cauchy-Kovalevskaya theorem was started in M. Nagumo's paper [44]. It is based on the fact that the initial value problem (0.3), (0.2) is equivalent to the integro-differential equation

$$u(t) = u_0 + \int_0^t f(\tau, u(\tau), Du(\tau))d\tau. \qquad (0.5)$$

The crucial point of that approach is the so-called Nagumo lemma estimating the derivative of a holomorphic function near the boundary (see 8.1., lemma 2; see also the estimate (2.2) in section 2.1.).

8. Also immediately starting from the Nagumo lemma, W. Walter [78] recently proved the Cauchy-Kovalevskaya theorem by using a weighted supremum norm for functions depending on time in conical domains.

9. Since a related singular integral operator is not bounded within the space of continuous functions equipped with the supremum norm, W. Walter's weighted supremum norm cannot be applied to solving initial value problems with generalized analytic initial functions. Modifying W. Walter's norm by using a family of subdomains, one can define a variant applicable to Hölder-continuous functions. The use of that modified weighted supremum norm enables us to reduce initial value problems with generalized analytic initial functions to the contraction-mapping principle, too. The family of subdomains entering into the definition of the modified weighted supremum norm shows that W. Walter's method is closely connected with the method of scales of Banach spaces because that norm can be interpreted as a norm for mappings into a scale depending on time (cf. also section 8.3.).

10. Since the initial value problem (0.3), (0.2) is equivalent to the integro-differential equation (0.5) the methods under consideration can be applied to integro-differential equations, too.

1. INITIAL VALUE PROBLEMS IN BANACH SPACES

The goal of this introductory chapter is not only to solve initial value problems in Banach spaces but also to prove some of the statements necessary for the further constructions.

1.1. Functions valued in Banach spaces

Let B be a given Banach space equipped with the norm $\|\cdot\|$. Let $u = u(t)$ be a function defined on the interval $[a, b]$ on the real axis so that its value $u(t)$ belongs to B for every $t \in [a, b]$. Then $u = u(t)$ is called a function valued in the Banach space B. Analogously one can investigate functions defined on other sets instead of an interval. Later on we shall especially regard functions defined on semi-open intervals $[0, T)$.

The function $u = u(t)$ valued in the Banach space B with the norm

$\|\cdot\|$ is said to be continuous at t_0 if the following condition holds:

For every $\varepsilon > 0$ there exists $\delta(\varepsilon) > 0$ such that $\|u(t) - u(t_0)\| < \varepsilon$ if $|t - t_0| < \delta(\varepsilon)$.

A function continuous at each point of its domain is called continuous.

1.2. Differentiation of functions valued in Banach spaces

Let $u = u(t)$ be a given function valued in the Banach space B. Suppose that the given function $u = u(t)$ satisfies the following condition at the point t_0:

For given $\varepsilon > 0$ there exist a number $\delta(\varepsilon) > 0$ and an element $a \in B$ such that

$$\left\|\frac{1}{t - t_0}(u(t) - u(t_0)) - a\right\| < \varepsilon \tag{1.1}$$

if $|t - t_0| < \delta(\varepsilon)$, $t \neq t_0$.

Then $u = u(t)$ is said to be differentiable at t_0. The element a is called the derivative of $u = u(t)$ at t_0 and is denoted by $\frac{du}{dt}(t_0)$. Notice that the difference quotient in (1.1) exists since B has a linear structure. It may be added that the derivative of a differentiable function is uniquely determined because there is at most one element a fulfilling the condition (1.1). From the definition of the derivative one gets easily that the usual rule for the derivative of a linear combination holds also in the case of functions valued in Banach spaces.

The following theorem holds also in the case of differentiable functions valued in Banach spaces:

Theorem. Suppose that the derivative of $u = u(t)$ vanishes identically in $[a, b]$, i.e.,

$$\frac{du}{dt}(t_0) = 0 \quad \text{for each} \quad t_0 \in [a, b]. \tag{1.2}$$

Then the function $u = u(t)$ must be constant in the whole interval $[a, b]$.

Proof. First, notice that the definition (1.1) and the assumption (1.2) yield immediately the estimate

$$\|u(t) - u(t_0)\| \leq \varepsilon|t - t_0| \tag{1.3}$$

if $|t - t_0| < \delta(\varepsilon)$. Second, we shall prove that (1.2) implies

$$\|u(t) - u(a)\| \leq \varepsilon|t - a| \tag{1.4}$$

for each $\varepsilon > 0$ and each $t \in [a, b]$. Otherwise there would exist a positive number ε_0 such that (1.4) with ε_0 instead of ε does not hold for each $t \in [a, b]$. Thus the set of all points $t' \in [a, b]$ fulfilling the contrary inequality

$$\|u(t') - u(a)\| > \varepsilon_0 |t' - a| \tag{1.5}$$

is not empty. Let t_* be the infimum of the set of all t'. Hence (1.4) holds with ε_0 for each $t \in [a, t_*)$, whereas in each neighbourhood of t_* there exists at least one point t' (with $t' > t_*$) fulfilling (1.5). Since $t \to t_*$ implies $u(t) \to u(t_*)$, the inequality (1.4) holds for $t = t_*$ and $\varepsilon = \varepsilon_0$, too. Thus all t' are larger than t_*.

Applying the triangle inequality, one gets

$$\|u(t') - u(a)\| \leq \|u(t') - u(t_*)\| + \|u(t_*) - u(a)\|.$$

In view of (1.3) the first term on the right-hand side can be estimated by $\varepsilon_0|t' - t_*|$ if the distance of t' and t_* is sufficiently small. The second term can be estimated by $\varepsilon_0|t_* - a|$ since (1.4) is valid for $t = t_*$, too. Altogether we get the inequality

$$\|u(t') - u(a)\| \leq \varepsilon_0|t' - t_*| + \varepsilon_0|t_* - a| = \varepsilon_0|t' - a|$$

contradicting (1.5). Therefore (1.4) holds for each $\varepsilon > 0$. Carrying out the limiting process $\varepsilon \to 0$, we get from (1.4), finally, that $u(t) = u(a)$ for each $t \in [a, b]$. This completes the proof of the theorem.

1.3. Integration of functions valued in Banach spaces

Let the function $u = u(t)$ valued in the Banach space B be defined and continuous in the interval $[a, b]$. Regard a sequence of subdivisions of the interval $[a, b]$ into a finite number of subintervals. Suppose that the k-th subdivision is given by the points $\tau_i^{(k)}$, $0 \leq i \leq n_k$, where

$$a = \tau_0^{(k)} < \tau_1^{(k)} < \dots < \tau_{n_k}^{(k)} = b.$$

Then define the so-called Cauchy-Riemann sums

$$s^{(k)} = \sum_{i=k}^{n_k} (\tau_i^{(k)} - \tau_{i-1}^{(k)}) u(\tau_{i-1}^{(k)}) \tag{1.6}$$

(which are defined in view of the linear structure of the Banach space B). Then the <u>integral</u>

$$\int_a^b d\tau \cdot u(\tau)$$

is defined as limit of the Cauchy-Riemann sums if the length of the longest subinterval of each subdivision tends to zero. Using the uniform continuity of a continuous function, it is easy to check that this limit exists, and that the integral does not depend on the method of subdividing the interval $[a, b]$. The proofs of these statements are identical with those in the case of real-valued functions.

Finally we define

$$\int_a^b \ldots = -\int_b^a \ldots$$

if $a > b$ and

$$\int_a^a \ldots = 0.$$

1.4. Properties of the integral

In the case of functions valued in Banach spaces (cf. 1.3.) the (definite) integral is defined as limit of a sequence of Cauchy-Riemann sums in the same way as in the case of real-valued functions. Thus in both cases many rules hold likewise. The integral of a linear combination, for instance, is equal to the corresponding linear combination of the integrals, i.e.,

$$\int_a^b d\tau\cdot(\alpha_1 u_1(\tau) + \alpha_2 u_2(\tau))$$
$$= \alpha_1 \int_a^b d\tau\cdot u_1(\tau) + \alpha_2 \int_a^b d\tau\cdot u_2(\tau).$$

Similarly we have

$$\int_a^b \ldots = \int_a^c \ldots + \int_c^b \ldots$$

if $a < c < b$ and

$$\int_a^b d\tau\cdot u_0 = (b - a)u_0$$

if the integrand is a constant element u_0 of the Banach space.

In this section we are going to prove only two statements.

First we estimate the norm of the integral

$$\int_a^b d\tau\cdot u(\tau).$$

This integral is the limit of the Cauchy-Riemann sums (1.6). From (1.6) we immediately get the estimate

$$\|s^{(k)}\| \leqq \sum_{i=1}^{n_k} (\tau_i^{(k)} - \tau_{i-1}^{(k)})\|u(\tau_i^{(k)})\|. \tag{1.7}$$

Taking into account that $\|u(\tau)\|$ depends continuously on τ, the right-hand sides of the last inequalities converge to the integral of $\|u(\tau)\|$ over the interval $[a, b]$. Further the left-hand sides of this inequalities converge to the norm of the integral of $u(\tau)$ and, consequently, (1.7) yields the following estimate:

<u>Theorem 1.</u> The integral of $u = u(\tau)$ can be estimated by

$$\left\|\int_a^b d\tau\cdot u(\tau)\right\| \leqq \int_a^b \|u(\tau)\|\,d\tau.$$

Now regard the indefinite integral

$$U(t) = \int_a^t d\tau\cdot u(\tau), \tag{1.8}$$

where $u = u(\tau)$ is again a continuous function defined in $[a, b]$. Applying the rules formulated above, one obtains easily $(t \neq t_0)$

$$\frac{1}{t-t_0}(U(t) - U(t_0)) - u(t_0)$$

$$= \frac{1}{t-t_0} \int_{t_0}^{t} d\tau \cdot (u(\tau) - u(\tau_0)).$$

Since $u = u(\tau)$ is continuous we have

$$\|u(\tau) - u(t_0)\| < \varepsilon$$

if $|t - t_0| < \delta(\varepsilon)$ and τ lies between t_0 and t.

Applying theorem 1 we get, consequently, the estimate

$$\left\| \frac{1}{t-t_0}(U(t) - U(t_0)) - u(t_0) \right\| < \varepsilon$$

if $|t - t_0| < \delta(\varepsilon)$. Thus the following theorem has been proved:

Theorem 2. The function $U = U(t)$ defined by (1.8) is differentiable and its derivative is given by

$$\frac{dU}{dt}(t_0) = u(t_0)$$

at every point t_0 of $[a, b]$.

According to the theorem of section 1.2. two functions with the same derivative differ in a constant. Consequently theorem 2 yields the following

Corollary. Let $u = u(\tau)$ be given. Then all functions $U = U(\tau)$ fulfilling the differential equation

$$\frac{dU}{dt} = u \tag{1.9}$$

are representable in the form

$$U(t) = u_0 + \int_a^t d\tau \cdot u(\tau),$$

where u_0 is the initial value $U(a)$. Consequently, the initial value problem $U(a) = u_0$ for the differential equation (1.9) is proved to be uniquely solvable, especially.

1.5. Solution of initial value problems in Banach spaces

Again assume that B is a given Banach space with the norm $\|\cdot\|$. Let, further, u_0 be a given element belonging to B. Then regard the initial value problem

$$\frac{du}{dt} = f(t,u), \tag{1.10}$$

$$u(0) = u_0 \tag{1.11}$$

in the Banach space B, i. e., we look for a function $u = u(t)$ valued in the Banach space B and defined in some t-interval for which the differential equation (1.10) as well as the initial condition (1.11) are fulfilled. Suppose that the right-hand side f(t,u) of the differential equation (1.10) satisfies the following conditions:

(I) There exist positive numbers R and T such that the right-hand side f(t,u) defines a continuous mapping of

$$\{t : 0 \leq t \leq T\} \times \{u \in B : \|u - u_0\| \leq R\}$$

into B.

(II) For each t we have

$$\|f(t,u_0)\| \leq K,$$

where K is a positive constant.

(III) There exists a positive constant L such that

$$\|f(t,u) - f(t,v)\| \leq L\|u - v\|,$$

where L is independent of t, u, and v (Lipschitz condition).

In view of the corollary from the section 1.4. the initial value problem (1.10), (1.11) is equivalent to the integral equation

$$u(t) = u_0 + \int_0^t d\tau \cdot f(\tau, u(\tau)) \tag{1.12}$$

in the Banach space B. Successively substituting into the right-hand side of (1.12), we define the approximations $u_k = u_k(t)$, k = 1, 2, ..., in the following way:

$$u_1(t) = u_0 + \int_0^t d\tau \cdot f(\tau, u_0), \tag{1.13}$$

$$u_{k+1}(t) = u_0 + \int_0^t d\tau \cdot f(\tau, u_k(\tau)) \tag{1.14}$$

(method of successive approximations). Using theorem 1 from section 1.4., the definition (1.13) and assumption (II) yield the estimate

$$\|u_1(t) - u_0\| \leq Kt. \tag{1.15}$$

The k-th iteration u_k must satisfy the condition

$$\|u_k(t) - u_0\| \leq R$$

in order to ensure the existence of the (k+1)-th iteration u_{k+1}. Since

$$\|u_k(t) - u_0\| \leq \|u_k(t) - u_{k-1}(t)\| + \ldots + \|u_1(t) - u_0\|$$

the last condition is fulfilled if

$$\|u_k(t) - u_{k-1}(t)\| \leq \varepsilon_k R, \tag{1.16}$$

where the ε_k are positive numbers satisfying the condition

$$\sum_{k=1}^{\infty} \varepsilon_k \leq 1.$$

Notice that the sequence $\varepsilon_1, \varepsilon_2, \ldots$ is avoidable. We use it, however, because the proof of the theorem in 3.3. is based on an analogous construction.

By virtue of (1.15) the inequality (1.16) is satisfied in the case $k = 1$ if t is restricted to the interval

$$0 \leq t \leq \frac{\varepsilon_1 R}{K}. \tag{1.17}$$

Taking into consideration the definition (1.14) for k and $k - 1$ as well as the Lipschitz condition (III), we get

$$\begin{aligned} &\| u_{k+1}(t) - u_k(t) \| \\ &\leq \int_0^t \| F(\tau, u_k(\tau)) - F(\tau, u_{k-1}(\tau)) \| \, d\tau \\ &\leq \int_0^t L \| u_k(\tau) - u_{k-1}(\tau) \| \, d\tau. \end{aligned}$$

Suppose that (1.16) is fulfilled for a fixed k. Then the last estimate shows that (1.16) holds also for $k + 1$ if

$$Lt\varepsilon_k \leq \varepsilon_{k+1}. \tag{1.18}$$

Thus the estimate (1.16) is proved by induction for every k provided t satisfies the inequalities (1.17) and (1.18), $k = 1, 2, \ldots$ It remains to prove that there exists an interval of positive length such that for all t belonging to this interval the inequality (1.17) and all inequalities (1.18) are satisfied at the same time. To this end we take

$$\varepsilon_{k+1} = Lt\varepsilon_k.$$

Therefore $\sum_{k=1}^{\infty} \varepsilon_k$ converges if

$$Lt < 1, \tag{1.19}$$

and we get $\sum_{k=1}^{\infty} \varepsilon_k = 1$ if we choose

$$\varepsilon_1 = 1 - Lt.$$

In this case condition (1.17) is equivalent to

$$0 \leq t \leq \frac{R}{LR + K}.$$

Hence we have proved:

If t varies in the interval

$$0 \leq t \leq \min\left(T, \frac{R}{LR + K}, \frac{1}{L}\right), \tag{1.20}$$

then $u_k(t)$ is defined for each k. On the other hand, in view of (1.16) the u_k form a Cauchy sequence for every t belonging to the interval (1.20). Therefore, the limit $u = u(t)$ exists for each such t. The convergence is uniform since the right-hand sides of (1.16) do not depend on t. Once more applying the Lipschitz condition (III), one obtains the estimate

$$\left\| \int_0^t d\tau \cdot \big(f(\tau, u_k(\tau)) - f(\tau, u(\tau))\big) \right\|$$
$$\leq L \cdot \sup_\tau \| u_k(\tau) - u(\tau) \| \cdot t.$$

Hence the right-hand sides of (1.14) converge to the right-hand side of (1.12), and the limit function $u = u(t)$ is proved to be a solution of the integral equation (1.12) and, consequently, of the initial value problem (1.10), (1.11). Summarizing the above arguments, we have proved the following

Theorem. Suppose that the right-hand side of the differential equation (1.10) fulfils the conditions (I), (II), and (III). Then there exists a solution $u = u(t)$ of the initial value problem (1.10), (1.11) in the interval (1.20). The solution can be obtained by successive approximations.

1.6. Uniqueness of the solution of initial value problems

Let u_1 and u_2 be two solutions of the same initial value problem (1.10), (1.11) in some t-interval $[0, t_*)$. Suppose that the right-hand side $f(t,u)$ of the differential equation (1.10) satisfies the conditions (I), (II), and (III) of 1.5. Since the initial value problem is equivalent to the integral equation (1.12), the difference $u_1 - u_2$ of the two given solutions u_1, u_2 is a solution of the equation

$$u_1(t) - u_2(t) = \int_0^t d\tau \cdot \big(f(\tau, u_1(\tau)) - f(\tau, u_2(\tau))\big).$$

In view of theorem 1 of 1.4. and condition (III) of 1.5. one obtains the estimate

$$\| u_1(t) - u_2(t) \| \leq Lt \sup_{0 \leq \tau \leq t} \| u_1(\tau) - u_2(\tau) \|$$
$$\leq Lt_* \sup_{0 \leq \tau \leq t} \| u_1(\tau) - u_2(\tau) \|$$

because $t \in [0, t_*)$. Provided that $Lt_* < 1$, the last inequality holds only if the supremum vanishes, i. e., the two solutions u_1 and u_2 must coincide.

Subdividing an interval of arbitrary length t_* into subintervals whose lengths are smaller than $\frac{1}{L}$, it follows that the same statement is valid without any restriction of t_*.

1.7. Initial value problems in the case of infinite systems of ordinary differential equations

Regard the infinite system of ordinary differential equations

$$\frac{du_k}{dt} = f_k(t, u_1, u_2, \ldots) \tag{1.21}$$

for infinitely many unknown real-valued functions $u_1, u_2, \ldots$ We look for a solution of (1.21) fulfilling the initial condition

$$u_k(0) = u_{ok}, \quad k = 1, 2, \ldots \tag{1.22}$$

We shall reduce the initial value problem (1.21), (1.22) to the initial value problem for differential equations in Banach spaces (that had been solved in section 1.5.). To this end we must interpret both the unknown vector $(u_1(t), \ldots)$ for each t and the initial vector $(u_{o1}, u_{o2}, \ldots)$ as elements of a suitably chosen Banach space B. The choice of this Banach space influences the sufficient conditions under which the solution of the initial value problem exists and is uniquely determined. We shall look for solutions $u = (u_1, u_2, \ldots)$ of (1.21), (1.22) that belong to $B = l_1$ for each t. The space l_1 consists of all vectors $(u_1, u_2, \ldots)$ with (real) components u_j such that

$$\sum_{j=1}^{\infty} |u_j| < +\infty.$$

The space l_1 turns out to be a Banach space if the norm of the infinite vector $u = (u_1, u_2, \ldots)$ is defined by

$$\|u\| = \sum_{j=1}^{\infty} |u_j|.$$

Now suppose that the right-hand sides $f_j(t, u_1, u_2, \ldots)$ satisfy the following conditions:

a) There exist positive numbers T and R such that each $f_j(t, u_1, u_2, \ldots)$ is defined and real-valued if $0 \leq t \leq T$ and $|u_k - u_{ok}| \leq R$, $k = 1, 2, \ldots$

b) There exists a positive constant K such that

$$\sum_{j=1}^{\infty} |f_j(t, u_{o1}, u_{o2}, \ldots)| \leq K$$

for each t, $0 \leq t \leq T$.

c) Every f_j satisfies a <u>Lipschitz condition</u>

$$|f_j(t_1, u_1, u_2, \ldots) - f_j(t_2, v_1, v_2, \ldots)|$$

$$\leq L_{jo}|t_1 - t_2| + \sum_{k=1}^{\infty} L_{jk}|u_j - v_j|,$$

where the series

$$\sum_{j=1}^{\infty} L_{jk}, \quad k = 0, 1, 2, \ldots$$

are assumed to be convergent and uniformly bounded, i. e., we suppose the existence of a further constant L such that

$$\sum_{j=1}^{\infty} L_{jk} \leq L \text{ for each } k = 0, 1, 2, \ldots$$ [1]).

In order to apply the theorem of 1.5. we must verify that the assumptions (I), (II), and (III) are fulfilled. First note that in view of b) the vector $f(t, u_0)$ belongs to $B = l_1$, where f means the vector $(f_1, f_2, \ldots)$ that is formed by the right-hand sides f_j of the differential equations (1.21). Immediately one gets from b) that assumption (II) is satisfied with the same K.

Take any $u \in l_1$ with $\|u - u_0\| \leq R$ where u_0 is the initial vector $(u_{01}, u_{02}, \ldots)$ that is assumed to belong to l_1. Since $\|u - u_0\| \leq R$ implies $|u_k - u_{ok}| \leq R$ for each k, the right-hand sides $f_j(t,u_1,u_2,\ldots)$ are defined in

$$\{t : 0 \leq t \leq T\} \times \{u \in l_1 : \|u - u_0\| \leq R\}. \tag{1.23}$$

By virtue of c) we have

$$\|f(t_1,u) - f(t_2,v)\| \leq L(|t_1 - t_2| + \|u - v\|)$$

for any two points (t_1, u) and (t_2, v) belonging to the cylinder (1.23). Thus the vector f(t, u) defines a continuous mapping of (1.23) into l_1 for which condition (I) is satisfied. Substituting $t_1 = t_2 = t$ into the last estimate, one gets further that the third condition (III) is satisfied, too.

Hence the theorem of 1.5. yields the following statement:

Suppose that the right-hand sides of the differential equations (1.21) satisfy the assumptions a), b), and c). Suppose further that the initial vector $u_0 = (u_{01}, u_{02}, \ldots)$ belongs to l_1. Then the initial value problem (1.21), (1,22) is solvable in the interval

$$0 \leq t \leq \min(T, \frac{R}{LR + K}, \frac{1}{L}).$$

In view of 1.6. the solution is uniquely determined in l_1.

---

[1]) The Lipschitz continuity of the f_j with respect to t and the convergence of the series

$$\sum_{j=1}^{\infty} L_{jo}$$

may be replaced by the weaker condition that the vector $(f_1, f_2, \ldots)$ depends continuously on t for every fixed $(u_1, u_2, \ldots)$. It may be added that the initial value problem (1.21), (1.22) is even solvable if the right-hand sides f_j are only measurable functions in t (cf. A. N. Tikhonov [60], K.G.Valeev and O.A. Zhautykov [76]).

2. SCALES OF BANACH SPACES

Assume that the right-hand side f of the differential equation (0.1) does depend on certain derivatives Du of the function u looked for, i. e., a partial differential equation of type (0.3) is to be investigated instead of an ordinary differential equation (0.1). Differential operators D entering into the right-hand side f of (0.3) are not bounded, in general. That is why in a fixed Banach space the initial value problem (0.3), (0.2) cannot be solved by using the method of successive approximations regarded in section 1.5.

On the other hand, differential operators may be interpreted as bounded operators if they are regarded not in a fixed Banach space B but in a suitably chosen family B_s of Banach spaces. An important example for such interpretation is given by holomorphic functions since their derivatives may be estimated by the Cauchy integral formula. Starting with an estimation of the derivatives of holomorphic functions in subsets (cf. 2.1.), in section 2.2. we will elaborate the concept of scales of Banach spaces in which initial value problems in the case of partial differential equations can be solved by the method of successive approximations. The basic idea of this approach is the following: A partial differential equation can be replaced by an operator equation with bounded operators mapping a scale of Banach spaces into itself.

2.1. The behaviour of the derivative of a holomorphic function in compact subsets

Let G be a bounded domain in the z-plane. Let, further, w = w(z) be a complex-valued function defined and continuous in $\overline{G}$ and holomorphic in G. Denote by H the space of all functions w = w(z) possessing these properties. The space H equipped with the supremum norm (maximum norm)

$$\|w\| = \sup_G |w(z)| \tag{2.1}$$

turns out to be a normed space. The limit function of a Cauchy sequence of functions belonging to H is continuous since convergence with respect to the supremum norm means uniform convergence. In view of Weierstrass' convergence theorem, moreover, the limit function of a uniformly convergent sequence of holomorphic functions is holomorphic, too. Thus the space H is proved to be complete, i.e., H is a Banach space.

Now let G' be a further domain whose closure $\overline{G'}$ is contained in the given domain G. The closure $\overline{G'}$ is, consequently, a compact subset of G. Denote by δ the (positive) distance of G' from the boundary. Then every closed disk centred at a point z_0 of G' belongs entirely to G if its radius r is smaller than δ (see the figure). Applying the Cauchy integral formula, the value of the derivative $\frac{dw}{dz}$ at the point z_0 may be represented by the integral

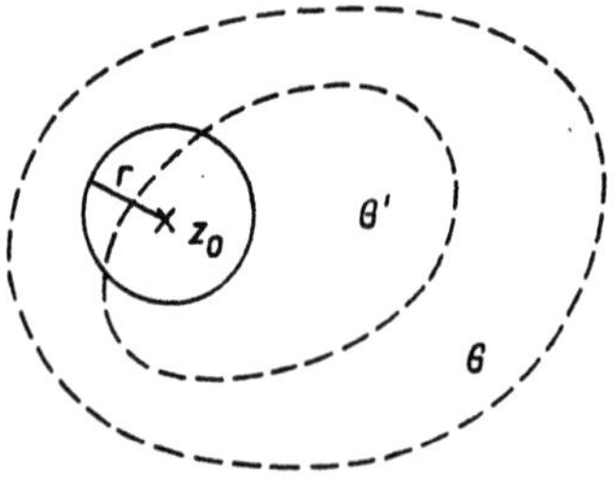

$$\frac{dw}{dz}(z_0) = \frac{1}{2\pi i} \int_{|z-z_0|=r} \frac{w(z)}{(z-z_0)^2} dz.$$

Since the definition of the norm in H implies that $|w(z)| \leqq \|w\|$ everywhere in G, one obtains the estimate

$$\left|\frac{dw}{dz}(z_0)\right| \leqq \frac{1}{2\pi} \frac{\|w\|}{r^2} 2\pi r = \frac{1}{r}\|w\|.$$

This inequality holds for each $r < \delta$. If r tends to δ, one gets for each z_0 belonging to G'

$$\left|\frac{dw}{dz}(z_0)\right| \leqq \frac{1}{\delta}\|w\|. \tag{2.2}$$

In the same way as H corresponds to G, the space H' is defined as the space of all complex-valued functions holomorphic in G' and continuous in $\overline{G'}$. Then the space H' equipped with the supremum norm

$$\|w\|_{G'} = \sup_{G'}|w(z)| \tag{2.3}$$

is also a Banach space. In order to distinguish the norm (2.1) in H from the norm (2.3) in H', the norm in H will be denoted by $\|\cdot\|_G$ instead of $\|\cdot\|$ if necessary. Using these notations, one obtains from (2.2) the estimate

$$\left\|\frac{dw}{dz}\right\|_{G'} \leqq \frac{1}{\delta}\|w\|_G$$

that proves the following theorem:

<u>Theorem.</u> The complex differentiation is a bounded operator mapping H into H' whose norm can be estimated by

$$\left\|\frac{d}{dz}\right\| \leqq \frac{1}{\delta}. \tag{2.4}$$

2.2. Definition of scales of Banach spaces

Regard once more the two Banach spaces H and H' introduced in section 2.1. Not only the complex differentiation $\frac{d}{dz}$ may be interpreted as bounded linear operator mapping H into H' but also the restriction to G' of a function $w = w(z)$ belonging to H may be interpreted as such operator, too. Denote this restriction by I. Then Iw is holomorphic in

G' and continuous in $\overline{G'}$ if $w = w(z)$ belongs to H. Equipping both H and H' with the supremum norm, we get the estimate

$$\|Iw\|_{G'} = \sup_{G'}|w(z)| \leq \sup_{G}|w(z)| \leq \|w\|_G,$$

i. e., I is proved to be an operator whose norm is not greater than 1,

$$\|I\| \leq 1.$$

We know, further, the following property of a holomorphic function:

If a holomorphic function defined in a domain G vanishes identically in a subdomain G', then the given holomorphic function must vanish identically in the whole domain G. Therefore, two holomorphic functions must be identical in the whole domain G if they coincide in a subdomain G'. The restrictions Iw_1 and Iw_2 to G' of two holomorphic functions w_1 and w_2 defined in the whole domain G are identical, consequently, if and only if $w_1 = w_2$ in the whole domain G. Thus the equality $Iw_1 = Iw_2$ of the restrictions to G' implies $w_1 = w_2$ in the whole domain G. Notice that an operator is said to be injective if two different elements possess images different from each other. Hence we can say that the restriction I mapping H into H' is injective. Summarizing these considerations, we have proved the following

Lemma. The restriction I has the following three properties:

a) I is linear.

b) I is a bounded operator whose norm is not greater than 1.

c) I is injective.

In the following we will regard not only two domains G and G' but a whole family of domains G_s where s is a real parameter varying in an interval $0 < s < s_0$, where s_0 is a given finite number. Strictly speaking, if G is a given bounded domain in the z-plane, we choose a family of subdomains G_s, $0 < s < s_0$, satisfying the following conditions:

a) the closure $G_{s'}$ is a compact subset of G_s if only $s' < s$,

b) the distance of $G_{s'}$ from the boundary ∂G_s of G_s can be estimated by

$$\text{dist } (G_{s'}, \partial G_s) \geq \text{const } (s - s'), \qquad (2.5)$$

where the constant is independent of s' and s, $s' < s$,

c) every point of G is contained in G_s for sufficiently large s.

Example. Let G be a disk centred at z_0 with radius r. Then a family G_s of subdomains, where s varies in the interval $0 < s < 1$, is given by

$$G_s = \{z : |z - z_0| < sr\}.$$

This family satisfies the three conditions formulated above. The constant entering into (2.5) is equal to r.

For any number s belonging to the interval $0 < s < s_0$ we define the space H_s of all complex-valued functions defined and continuous in $\overline{G}_s$ and holomorphic in G_s. The space H_s equipped with the supremum norm

$$\|w\|_s = \sup_{G_s} |w(z)|$$

again turns out to be a Banach space. Denote the restriction of a function $w \in H_s$ to $G_{s'}$, where $s' < s$ by $I_{s,s'}$. In view of the lemma formulated above the operator $I_{s,s'}$ is linear and injective and its norm is not greater than 1.

For short we say that a Banach space B is injected into another Banach space B' if there exists a linear and injective operator I with $\|I\| \leqq 1$ mapping B into B'. This operator allows us to interpret B as a subset of B'. Indeed, I defines a 1-1-mapping between B and IB since I is injective. The assumption $\|I\| \leqq 1$ implies, further, that the norm in B is not smaller than the norm induced on B by B'.

Using this definition of an injection, we can say that the Banach spaces H_s, $0 < s < s_0$, form a family, where each H_s is injected into each space $H_{s'}$ with $s' < s$.

Finally we introduce the concept of a scale of Banach spaces. A family of Banach spaces B_s, where s varies in the open interval $0 < s < s_0$, is called a scale of Banach spaces if every B_s is injected into each space $B_{s'}$ with a smaller s'. If a scale of Banach spaces is given, then we have not only a family of Banach spaces B_s, $0 < s < s_0$, but at the same time also a family of linear and injective operators $I_{s,s'}$ with the norms $\|I_{s,s'}\| \leqq 1$ mapping B_s into $B_{s'}$, $0 < s' < s < s_0$.

The family H_s is a special scale of Banach spaces. Later on we shall construct scales of Banach spaces whose elements are generalized analytic functions.

2.3. Generalized Cauchy-Riemann operators in scales of Banach spaces

Let us start again with the pair H, H' of Banach spaces of functions that are holomorphic in G and G' resp. and continuous in $\overline{G}$ and $\overline{G'}$ resp. (cf. 2.1.). Then in view of the theorem of 2.1. the complex differentiation $\frac{d}{dz}$ is a bounded operator mapping H into H' where its norm can be estimated by (2.4).

Applying this theorem to the scale H_s of Banach spaces introduced in section 2.2., one obtains the following statement:

For any pair s', s with $s' < s$ the complex differentiation $\frac{d}{dz}$ maps H_s into $H_{s'}$, where the norm of this mapping can be estimated by

$$\left\|\frac{d}{dz}\right\| \leq \frac{\text{const}}{s - s'} .$$

It may be added that the constant in the last inequality is the reciprocal of the constant entering into (2.5).

In a similar way as we generalized the scale H_s to an arbitrary scale of Banach spaces B_s in the previous section 2.2., we now intend to generalize the operator $\frac{d}{dz}$ to the case of an arbitrary scale. As a suitable generalization of $\frac{d}{dz}$ we consider operators with the following property:

For any pair s', s with s' < s the operator maps each B_s into each $B_{s'}$, where the norm of this mapping does not exceed the number

$$\frac{\text{const}}{s - s'} ,$$

where the constant does not depend on s' and s.

Since such operators generalize the complex differentiation of holomorphic functions we will denominate them as <u>generalized Cauchy-Riemann operators</u>. Examples different from the ordinary complex differentiation will be given by differential operators acting in spaces of generalized analytic functions (cf. chapter 7.; see also 9.1.1.).

2.4. Dual scales

First recall the concept of dual spaces. Let B be a given Banach space. Then its dual space B^* is defined as the space of all (real-valued) linear and bounded functionals of B. Norming the dual space B^* by the operator norm of the functionals belonging to B^*, the dual space B^* itself turns out to be a Banach space.

Second recall the definition of <u>adjoint operators</u>. For this end regard two Banach spaces B and B' and a linear and bounded operator T mapping B into B'. Denote an arbitrary element of B by x, whereas x' is an arbitrary element belonging to B'. Denote further the elements of the dual B^* of B by X and those of the dual B'^* of B' by X'. Then the adjoint operator T^* of T maps B'^* into B^*. It is defined by

$$(T^*X')[x] = X'(Tx), \tag{2.6}$$

where x is an arbitrary element of B. From this well-known definition it follows immediately that

$$\|(T^*X')[x]\| \leq \|X'\|\cdot\|Tx\| \leq \|X'\|\cdot\|T\|\cdot\|x\| ,$$

i.e.,

$$\|T^*X'\| \leq \|T\|\cdot\|X'\|$$

for each X' of B'^*. Thus T^* is proved to be a bounded operator, where

$$\|T^*\| \leq \|T\|. \tag{2.7}$$

It is obvious that T^* is linear, too.

Now let B_s, $0 < s < s_0$, be a given scale of Banach spaces. Consequently, there is also given a family of linear and injective operators $I_{s,s'}$ with the norms $\|I_{s,s'}\| \leq 1$, $0 < s' < s < s_0$. Regard further the family of the dual spaces B_s^*, $0 < s < s_0$, and the adjoint operators $I_{s,s'}^*$ mapping $B_{s'}^*$ into B_s^*, $0 < s' < s < s_0$. Obviously, all adjoint operators are linear and bounded, where in view of (2.7) the inequality

$$\|I_{s,s'}^*\| \leq \|I_{s,s'}\| \leq 1$$

holds. Unfortunately the adjoint operator $I_{s,s'}^*$ is not injective, in general. However, it is easy to formulate sufficient conditions under which $I_{s,s'}^*$ is in fact injective. The following lemma is valid:

Lemma. If B_s is dense in $B_{s'}$, then $I_{s,s'}^*$ is injective.

Proof. If

$$I_{s,s'}^* X_1' = I_{s,s'}^* X_2' \tag{2.8}$$

for two elements X_1', X_2' belonging to $B_{s'}^*$, then

$$(I_{s,s'}^* X_1')[x] = (I_{s,s'}^* X_2')[x]$$

for each element x of B_s. By virtue of (2.6) we get

$$X_1'(I_{s,s'}x) = X_2'(I_{s,s'}x) \tag{2.9}$$

for each x belonging to B_s. On the other hand, B_s is dense in $B_{s'}$. This means that for every element x' of $B_{s'}$ there exists a sequence of elements x_n in B_s such that

$$x' = \lim_{n\to\infty} I_{s,s'}x_n.$$

Thus (2.9) implies that

$$X_1'(x') = X_2'(x')$$

for each x' belonging to $B_{s'}$. The last equality means, however, that $X_1' = X_2'$. Thus the equation (2.8) is possible only if X_1' and X_2' coincide, i. e., the lemma has been proved.

Assume now that B_s is dense in $B_{s'}$ for any pair s,s' with $0 < s' < s < s_0$. Then the lemma shows that each $I_{s,s'}^*$ is an injective mapping of $B_{s'}^*$ into B_s^*. Denote s by $s_0 - \sigma'$ and s' by $s_0 - \sigma$. Then $0 < s' < s < s_0$ is equivalent to $0 < \sigma' < \sigma < s_0$. Denote further $B_s^* = B_{s_0-\sigma'}^*$ by $D_{\sigma'}$ and analogously, $B_{s'}^* = B_{s_0-\sigma}^*$ by D_σ. Thus $I_{s,s'}^* = I_{s_0-\sigma',s_0-\sigma}^*$ is a linear and injective operator with the norm $\|I_{s_0-\sigma',s_0-\sigma}\| \leq 1$ mapping D_σ into $D_{\sigma'}$. Consequently, the family $D_\sigma = B_{s_0-s'}^*$ of the dual spaces forms also a scale of Banach spaces. It is called the <u>dual scale</u> to the given scale. Summarizing the above arguments, the following theorem has been proved:

<u>Theorem.</u> Suppose that each B_s of a given scale of Banach spaces is dense in each $B_{s'}$, $0 < s' < s < s_0$. Then the dual spaces $B^*_{s_0-s'}$ form a scale of Banach spaces, too.

It remains to add an example for a scale of Banach spaces for which each B_s is dense in each $B_{s'}$, $0 < s' < s < 1$.

For this end we regard again the Banach spaces H_s defined in section 2.2., where G_s is the disk centred at $z = 0$ with radius sr, where r is a given positive number. Let $w = w(z)$ be an arbitrary element of $H_{s'}$. This means that $w = w(z)$ is holomorphic if $|z| < s'r$ and continuous if $|z| \leqq s'r$. Define a new function $\tilde{w}$ by $\tilde{w}(z) = w(qz)$, where q is a fixed number, $0 < q < 1$. Then $w = w(z)$ is holomorphic if $|z| < \frac{s'r}{q}$ and continuous if $|z| \leqq \frac{s'r}{q}$.

The distance of the points z and qz is equal to $|z - qz| = (1 - q)|z|$. Thus this distance may be estimated by $(1 - q)s'r$ if the point z belongs to the closed disk with radius $s'r$. On the other hand, the given function $w = w(z)$ is continuous and, therefore, uniformly continuous in this disk. Hence it follows

$$|w(z) - \tilde{w}(z)| = |w(z) - w(qz)| < \frac{\varepsilon}{2} \tag{2.10}$$

where ε is an arbitrary positive number, $|z| \leqq s'r$ and q is chosen nearly enough to 1. Since $\tilde{w} = \tilde{w}(z)$ is holomorphic if $|z| < \frac{s'r}{q}$ this function may be represented by its Taylor expansion in the open disk with radius $\frac{s'r}{q}$. This Taylor expansion converges uniformly in each closed subset. Thus there exists a polynomial $p(z)$ such that

$$|\tilde{w}(z) - p(z)| < \frac{\varepsilon}{2} \tag{2.11}$$

if $|z| \leqq s'r$, where $p(z)$ is the sum of a finite number of terms of the Taylor expansion of $w = w(z)$. From (2.10) and (2.11) we obtain

$$|w(z) - p(z)| < \varepsilon$$

for each z with $|z| \leqq s'r$ and, consequently,

$$\|w - p\| = \sup_{|z| \leqq s'r} |w(z) - p(z)| \leqq \varepsilon. \tag{2.12}$$

The last estimate proves that the set of all polynomials $p = p(z)$ is dense in $H_{s'}$. On the other hand each polynomial is holomorphic in the whole plane. Therefore all polynomials belong to H_s, $s > s'$. Thus the estimate (2.12) shows also that H_s is dense in $H_{s'}$ if $s' < s$.

3. SOLUTION OF INITIAL VALUE PROBLEMS IN SCALES OF BANACH SPACES

Whereas in chapter 1 we solved differential equations only in Banach spaces, we now intend to solve differential equations even in scales of Banach spaces.

3.1. Differential equations in scales of Banach spaces

In order to explain the concept of differential equations in scales of Banach spaces, we start with a simple but typical example. We look for a complex-valued function $w = w(t,z)$ depending on both a real variable t and a complex variable z. Assume that this function depends holomorphically on z for each t. Regard the partial differential equation

$$\frac{\partial w}{\partial t} = \frac{\partial w}{\partial z}, \tag{3.1}$$

where the differentiation $\frac{\partial}{\partial z}$ means the ordinary complex differentiation with respect to the variable z. Such interpretation is possible because $w = w(t,z)$ is supposed to be a holomorphic function in z. Now regard the scale H_s, $0 < s < s_0$, of Banach spaces of holomorphic functions introduced in section 2.2. Then the right-hand side of the differential equation may be interpreted as an operator mapping each H_s into $H_{s'}$, if $s' < s$. Since we look for solutions holomorphic in z for each t we may interpret the solution as an element of every Banach space H_s for each fixed t. Summarizing these interpretations of both the right-hand side and the solution of the differential equation (3.1), we get the following concept of differential equations in scales of Banach spaces:

Let B_s, $0<s<s_0$, be a given scale of Banach spaces. Let, further, $F(t,u)$ be an operator mapping each B_s into each $B_{s'}$, if $s' < s$ and t is fixed. Then we look for a function $u = u(t)$ belonging to each B_s for every fixed t and satisfying the differential equation

$$\frac{du}{dt} = F(t,u), \tag{3.2}$$

where the derivative $\frac{du}{dt}$ is defined in every Banach space B_s of the given scale in the sense of section 1.2. If we interpret $u(t)$ as an element of B_s , then $F(t,u)$ belongs only to $B_{s'}$. Therefore, also $\frac{du}{dt}(t)$ must be interpreted as element of $B_{s'}$. Properly speaking, the left-hand side $\frac{du}{dt}$ of (3.2) must be replaced by $I_{s,s'} \frac{du}{dt}$, where $I_{s,s'}$ is the injective operator connecting B_s with $B_{s'}$. For short we will use, however, the simplified denotation $\frac{du}{dt}$.

Finally we would like to point out that the interpretation of the right-hand side as an operator $F(t,u)$ (mapping a scale B_s into itself) replaces the right-hand sides $f(t,u,Du)$ of (0.3) (depending on further differentiations).

In section 3.5. we shall solve the initial value problem for the differential equation (3.2), i.e., we shall look for a solution of this differential equation satisfying the additional condition

$$u(0) = u_0, \tag{3.3}$$

where u_0 is a given element belonging to every Banach space of the given scale B_s.

3.2. Some preliminaries

In order to solve the initial value problem (3.2), (3.3), the right side $F(t,u)$ must satisfy some conditions. These conditions are similar to those for the right-hand sides of the initial value problems (1.10), (1.11) in Banach spaces (conditions (I), (II), and (III) in 1.5.). In order to deduce the analogous conditions in the case of initial value problems (3.2), (3.3) in scales of Banach spaces, we shall again take a simple partial differential equation as an example.

Let G be the disk centred at $z=0$ with radius r. Let, moreover, G_s be the disk with radius sr, $0 < s < 1$, and with the same centre $z = 0$. The space of all complex-valued functions holomorphic in G_s and continuous in $\overline{G_s}$ is denoted by H_s (cf. section 2.2.). Then the spaces H_s equipped with the supremum norm form a scale of Banach spaces.

Now regard the initial value problem

$$\frac{\partial w}{\partial t} = \frac{1}{2 - w}\,\frac{1}{1 - t}\,\frac{\partial w}{\partial z}, \tag{3.4}$$

$$w(0,z) = w_0(z), \tag{3.5}$$

where the initial function $w_0 = w_0(z)$ is supposed to be holomorphic in G and bounded by 1, i.e.,

$$|w_0(z)| \leq 1 \quad \text{if} \quad |z| < r.$$

For each s, $0 < s < 1$, the initial function $w_0 = w_0(z)$ belongs, consequently, to H_s and its norm in H_s may be estimated by

$$\|w_0\|_s \leq 1. \tag{3.6}$$

In the following we regard only such complex-valued functions $w = w(t,z)$ that depend holomorphically on z. Thus the differentiation $\frac{\partial w}{\partial z}$ in (3.4) may be understood as ordinary complex differentiation. If for any fixed t the function $w = w(t,z)$ is defined and continuous in $\overline{G_s}$ and holomorphic in G_s, then (for this fixed t) it may be interpreted as element of H_s. The whole function $w = w(t,z)$ may be interpreted, consequently, as a curve in H_s, where t is the parameter of this curve. Now regard, conversely, any element $w = w(z)$ of H_s (if a complex-valued function

$w = w(t,z)$ does not depend on t we write $w(z)$ instead of $w(t,z)$). Take any $w \in H_s$ with

$$\|w - w_0\|_s \leqq R,$$

where R is a fixed positive number. In view of (3.6) we get the estimate

$$|w(z)| \leqq \|w\|_s \leqq \|w - w_0\|_s + \|w_0\|_s \leqq R + 1 \tag{3.7}$$

for each z with $|z| \leqq sr$. Now restrict the number R to the interval $0 < R < 1$. Then one obtains

$$\left|\frac{1}{2 - w(z)}\right| \leqq \frac{1}{2 - (R + 1)} = \frac{1}{1 - R}. \tag{3.8}$$

Analogously we have

$$\left|\frac{1}{1 - t}\right| \leqq \frac{1}{1 - T} \tag{3.9}$$

if $0 \leqq t \leqq T$ and T is a fixed number with $0 < T < 1$. Especially, the right-hand side of the differential equation (3.4) is defined in the whole cylinder

$$\{t : 0 \leqq t \leqq T\} \times \{w \in H_s : \|w - w_0\|_s \leqq R\} \tag{3.10}$$

if $0 < T < 1$ and $0 < R < 1$. On the other hand, $\frac{\partial w}{\partial z}$ belongs to $H_{s'}$, and, therefore, the right side of (3.4) defines a mapping of the cylinder (3.10) into $H_{s'}$. This mapping depends continuously on t since in view of

$$\frac{1}{1 - t'} - \frac{1}{1 - t} = \frac{t' - t}{(1 - t')(1 - t)}$$

and (3.7) and (3.8) one gets

$$\left\|\frac{1}{2 - w}\,\frac{1}{1 - t'}\,\frac{\partial w}{\partial z} - \frac{1}{2 - w}\,\frac{1}{1 - t}\,\frac{\partial w}{\partial z}\right\|_{s'}$$

$$\leqq \frac{|t' - t|}{(1 - R)(1 - T)^2}\left\|\frac{\partial w}{\partial z}\right\|_{s'}$$

if $0 \leqq t \leqq T$ and $0 \leqq t' \leqq T$.

In order to investigate the dependence of the right-hand side of (3.4) on the function $w = w(z)$, we choose any two elements w_1 and w_2 of H_s satisfying the condition $\|w_j - w_0\|_s \leqq R$, $j = 1, 2$. Take any s', $0 < s' < s$. Let, moreover, z_0 be an arbitrary point belonging to $G_{s'}$. Cauchy's integral formula yields

$$\frac{\partial w_j}{\partial z}(z_0) = \frac{1}{2\pi i}\int_{|z-z_0|=\delta} \frac{w_j(z)}{(z-z_0)^2}dz, \tag{3.11}$$

where $0 < \delta < (s-s')r$, $j = 1, 2$. Since the disk centred at z_0 with radius δ is contained in G_s, from the formula (3.11) it follows, firstly,

$$\left|\frac{\partial w_j}{\partial z}(z_0)\right| \leqq \frac{1}{2\pi}\,\frac{1}{\delta^2}\|w_j\|_s\; 2\pi\delta \leqq \frac{\|w_j\|_s}{\delta} \leqq \frac{1 + R}{\delta},$$

where we took into consideration the estimate (3.7). Secondly, the formula (3.11) leads to the estimate

$$\left|\frac{\partial w_2}{\partial z}(z_0) - \frac{\partial w_1}{\partial z}(z_0)\right| \leq \frac{1}{2\pi}\frac{1}{\delta^2}\|w_2 - w_1\|_s 2\pi\delta = \frac{\|w_2 - w_1\|_s}{\delta} .$$

If δ tends to $(s - s')r$, from the two last estimates one gets

$$\left|\frac{\partial w_j}{\partial z}(z_0)\right| \leq \frac{1 + R}{(s - s')r} , \tag{3.12}$$

$$\left|\frac{\partial w_2}{\partial z}(z_0) - \frac{\partial w_1}{\partial z}(z_0)\right| \leq \frac{\|w_2 - w_1\|_s}{(s - s')r}. \tag{3.13}$$

On the other hand, we have

$$\frac{1}{2 - w_2}\frac{1}{1 - t}\frac{\partial w_2}{\partial z} - \frac{1}{2 - w_1}\frac{1}{1 - t}\frac{\partial w_1}{\partial z}$$

$$= \frac{1}{2 - w_2}\frac{1}{1 - t}\left(\frac{\partial w_2}{\partial z} - \frac{\partial w_1}{\partial z}\right) + \frac{w_2 - w_1}{(2 - w_2)(2 - w_1)}\frac{1}{1 - t}\frac{\partial w_1}{\partial z}.$$

Taking into account the estimates (3.8), (3.9), (3.12), and (3.13), we obtain, finally,

$$\left\|\frac{1}{2 - w_2}\frac{1}{1 - t}\frac{\partial w_2}{\partial z} - \frac{1}{2 - w_1}\frac{1}{1 - t}\frac{\partial w_1}{\partial z}\right\|_{s'} \tag{3.14}$$

$$\leq \frac{2}{(1 - R)^2(1 - T)r}\frac{1}{s - s'}\|w_2 - w_1\|_s$$

since z_0 is an arbitrary point of $G_{s'}$. Thus we have proved that the right-hand side of (3.4) is a generalized Cauchy-Riemann operator (cf. 2.3.) defined in the cylinder (3.10). The last estimate means also that the right-hand side of (3.4) depends continuously on w. On the other hand, we know already that the right-hand side of (3.4) depends continuously on t, too. Using the triangle inequality in $H_{s'}$, it is easy to check, consequently, that the right-hand side of (3.4) defines a continuous operator in the whole cylinder (3.10), i.e., the right-hand side is also a continuous function in (t,w).

At last we substitute the initial function $w_0 = w_0(z)$ into the right-hand side of (3.4). Regard an arbitrary point z_0 of G_j. Then the Cauchy integral formula (3.11) (with $j = 0$) and the assumption (3.6) yield

$$\left|\frac{\partial w_0}{\partial z}(z_0)\right| \leq \frac{1}{\delta},$$

where $0 < \delta < (1 - s)r$ (notice that the initial function $w_0 = w_0(z)$ is given in the whole domain G). Carrying out the limiting process $\delta \to (1 - s)r$, we obtain the estimate

$$\left\|\frac{\partial w_0}{\partial z}\right\|_s \leq \frac{1}{(1 - s)r} .$$

Once more taking into account the estimate (3.9), by virtue of

$$|2 - w_0(z)| \geq 1$$

one gets, finally, the estimate

$$\left\| \frac{1}{2 - w_0} \frac{1}{1 - t} \frac{\partial w_0}{\partial z} \right\|_s \leq \frac{1}{(1 - T)r} \frac{1}{1 - s}.$$

Summarizing these estimates, the following three properties of the right-hand side of (3.4) have been proved:

First it defines a continuous mapping of the cylinder (3.10) into H_s.

After the substitution $w = w_0$, second, the norm of the right-hand side in H_s can be estimated by

$$\frac{1}{(1 - T)r} \frac{1}{1 - s'} .$$

Third the right-hand side defines a generalized Cauchy-Riemann operator (cf. (3.14)).

Now we intend to investigate the general initial value problem (3.2), (3.3) instead of the specialized one (3.4), (3.5). Starting from the three properties of the right-hand side of (3.4) deduced above, we assume that the right-hand side F(t,u) of (3.4) satisfies the following three conditions, where B_s, $0 < s < s_0$, is a given scale of Banach spaces [1]):

(I) Firstly, assume that there exist positive numbers R and T such that for any pair s', s with $0 < s' < s < s_0$ the right-hand side F(t,u) defines a continuous mapping of

$$\{t : 0 \leq t \leq T\} \times \{u \in B_s : \|u - u_0\|_s \leq R\} \qquad (3.15)$$

into $B_{s'}$.

(II) Suppose, secondly, that the continuous function defined by $F(t,u_0)$ satisfies, with a positive constant K, the estimate

$$\|F(t,u_0)\|_s \leq \frac{K}{s_0 - s}$$

for each s, $0 < s < s_0$.

(III) Thirdly, we assume that F(t,u) is a generalized Cauchy-Riemann operator, i.e., there exists a positive constant C independent of t, u, v, s and s' such that

$$\|F(t,u) - F(t,v)\|_{s'} \leq \frac{C\|u - v\|_s}{s - s'} \qquad (3.16)$$

for any pair s', s with $0 < s' < s < s_0$ and for any pair u, v belonging to the cylinder (3.15).

[1]) See T. Nishida [47] .

In the next section we shall prove that the initial value problem (3.2), (3.3) is solvable under the conditions (I), (II), and (III).

To conclude the present section, let us additionally remark that condition (I) on F(t,u) may be weakened in the following way:

It is sufficient to suppose that F(t,u) depends continuously on the variable t. Indeed, applying the triangle inequality in $B_{s'}$, we have

$$\|F(t,u) - F(t',v)\|_{s'}$$
$$\leq \|F(t,u) - F(t',u)\|_{s'} + \|F(t',u) - F(t',v)\|_{s'}.$$

The first term of the right-hand side of the last inequality is arbitrarily small if $|t' - t|$ is sufficiently small. In view of (3.16) the second term is also arbitrarily small if $\|u - v\|_s$ is sufficiently small. Thus generalized Cauchy-Riemann operators are continuous in (t,u) if only they are partially continuous in t.

Analogously, the continuity of f(t,u) in condition (I) of section 1.5. may be replaced by the weaker assumption that f(t,u) depends continuously in t.

3.3. The method of successive approximations in scales of Banach spaces

Let B_s, $0 < s < s_0$, be a given scale of Banach spaces, where s_0 is a finite number. Then we regard the initial value problem (3.2), (3.3). Suppose that the right-hand side F(t,u) of (3.2) satisfies the three conditions (I), (II), and (III) formulated in the previous section 3.2. Our goal is to construct a solution u = u(t) belonging to each B_s for some t-interval. In view of the corollary of theorem 2 in section 1.4. the initial value problem (3.2), (3.3) is equivalent to the integral equation

$$u(t) = u_0 + \int_0^t d\tau \cdot F(\tau, u(\tau)) \tag{3.17}$$

in the scale B_s. If $u = u(\tau)$ belongs to B_s, then the left-hand side u(t) of (3.17) is an element of $B_{s'}$ with $s' < s$ because the right-hand side F(t,u) of (3.2) maps the cylinder (3.15) into $B_{s'}$. Properly speaking, the left-hand side u(t) of (3.15) must be replaced, consequently, by $I_{s,s'}u(t)$, where $I_{s,s'}$ is again the injective operator entering into the definition of a scale of Banach spaces (cf. section 2.2.). For short we use the simplified denotation u(t) for the left-hand side of (3.17).

The reduction of the initial value problem (3.2), (3.3) to the integral equation (3.17) in the scale B_s of Banach spaces, $0 < s < s_0$, is completely analogous to the reduction of the initial value problem (1.10), (1.11) to the integral equation (1.12) in a fixed Banach space B. Using the right-hand side of (3.17), we define now successive approximations, in the same way as done in section 1.5.:

The first approximation $u_1(t)$ is defined by

$$u_1(t) = u_0 + \int_0^t d\tau \cdot F(\tau, u_0), \tag{3.18}$$

whereas the further approximations $u_k(t)$ are defined by induction, namely

$$u_{k+1}(t) = u_0 + \int_0^t d\tau \cdot F(\tau,\ u_k(\tau)). \tag{3.19}$$

In accordance with condition (I) the integrand depends continuously on τ. The two last equations (3.18) and (3.19) correspond to the equations (1.13) and (1.14) in the case of differential equations in a fixed Banach space. The equations (3.18) and (3.19) show following difficulty in the case of differential equations in scales of Banach spaces:

Since $F(t,u)$ maps B_s into $B_{s'}$, $s' < s$, the (k+1)-th approximation u_{k+1} belongs only to $B_{s'}$ if the k-th approximation u_k belongs (for some t-interval) to B_s. First of all we have to construct, consequently, a common t-interval in which all $u_k(t)$ are defined and belong to each B_s. In this t-interval we shall carry out the limiting process $k \to \infty$. The limit function $u = u(t)$,

$$u(t) = \lim_{k\to\infty} u_k(t)$$

will turn out to be a solution of the integral equation (3.17) and, therefore, it will be a solution of the initial value problem (3.2), (3.3), too. We shall see, further, that the t-interval in question will depend on s. In order to get to know how the t-interval depends on the index s, we start with an estimate of the norm of $u_1 = u_1(t)$ in B_s, $0 < s < s_0$. Since u_0 belongs to every B_s, in view of condition (I) of 3.2. the element $F(t,u_0)$ belongs to every $B_{s'}$ if only t satisfies the inequality $0 \leq t \leq T$. Thus by virtue of the definition (3.18) the element $u_1(t)$ belongs to each $B_{s'}$ in the same t-interval. Replacing s' again by s and applying condition (II) of 3.2., one obtains, finally,

$$\|u_1(t) - u_0\|_s \leq \int_0^t \|F(\tau,u_0)\|_s d\tau \leq \frac{Kt}{s_0 - s}. \tag{3.20}$$

Restrictring t to the interval

$$0 \leq t < a(s_0 - s) \tag{3.21}$$

depending on s, where a is any positive number, we get the inequality

$$\|u_1(t) - u_0\|_s \leq Ka \tag{3.22}$$

whose right-hand side is independent on t and s. In order to ensure that the t-interval (3.21) is not greater than the original one, we must assume that $as_0 \leqq T$, i.e.,

$$a \leqq \frac{T}{s_0} . \qquad (3.23)$$

Restricting the semi-open interval (3.21) to the open one

$$0 < t < a(s_0 - s),$$

the inequality (3.20) may be rewritten as

$$\|u_1(t) - u_0\|_s \leqq \frac{Ka}{\frac{a(s_0-s)}{t}} < \frac{Ka}{\frac{a(s_0-s)}{t} - 1}$$

so that

$$\|u_1(t) - u_0\|_s \left(\frac{a(s_0-s)}{t} - 1\right) < Ka. \qquad (3.24)$$

Now assume that $u(t)$ is defined for each t with $0 \leqq t < a(s_0-s)$ and belongs to B_s if s is any number satisfying the inequality $0 < s < s_0$. Then we define the functional

$$M(u) = \sup_{0<s<s_0} \sup_{0<t<a(s_0-s)} \|u(t)\|_s \left(\frac{a(s_0-s)}{t} - 1\right)$$

that may be equal to $+\infty$. Provided that $M(u)$ is finite, from the definition of $M(u)$ one gets immediately the estimate

$$\|u(t)\|_s \leqq \frac{tM(u)}{a(s_0-s) - t} \qquad (3.25)$$

for each pair s, t satisfying the inequalities $0 < s < s_0$, $0 < t < a(s_0 - s)$. It may be added that the estimate (3.25) shows that $u(0)$ vanishes necessarily in each B_s if $M(u)$ is finite.

Now return to the element $u_1(t)$ defined by (3.18). Taking into account the estimate (3.24) being true for each s with $0 < s < s_0$ and for each t with $0 < t < a(s_0 - s)$, we see that $M(u_1 - u_0)$ is finite and may be estimated by

$$M(u_1 - u_0) \leqq Ka. \qquad (3.26)$$

We must be careful with the definition of the successive approximations u_k by formula (3.19). In order to define $u_{k+1}(t)$ we must ensure, first, that $u_k(t)$ satisfies the inequality

$$\|u_k(t) - u_0\|_s \leqq R \qquad (3.27)$$

because the integrand $F(\tau, u_k(\tau))$ of (3.19) is defined only in the cylinder (3.15). Second we must ensure the convergence of the u_k. Both conditions may be satisfied if the t-interval will be restricted in such a manner that

$$\|u_k(t) - u_{k-1}(t)\|_s \leqq \varepsilon_k R$$

for every $k = 1, 2, \ldots$, where ε_k are positive numbers satisfying the condition

$$\sum_{k=1}^{\infty} \varepsilon_k \leq 1. \tag{3.28}$$

By virtue of (3.20) the inequality (3.27) is satisfied already for $k = 1$ in the t-interval (3.21) if

$$a \leq \frac{\varepsilon_1 R}{K}. \tag{3.29}$$

We shall prove that (3.27) can be satisfied for each k if we diminish the t-interval step by step. By this diminution we must ensure, however, that the length of the limit interval is positive. The k-th approximation u_k will be defined if

$$0 \leq t < a_k(s_0 - s), \tag{3.30}$$

where $a_1 > a_2 > \ldots > a_k > 0$ and the number a (introduced above, cf. formula (3.21)) is denoted by a_1. Corresponding to the t-intervals (3.30) we must regard a sequence of funcionals

$$M_k(u) = \sup_{0<s<s_o} \; \sup_{0<t<a_k(s_0-s)} \|u(t)\|_s \left(\frac{a_k(s_0-s)}{t} - 1\right)$$

instead of the only functional M(u) defined above. Since $a = a_1$ from now on the functional M(u) is denoted by $M_1(u)$. From the definition of the functionals $M_k(u)$ one obtains immediately that

$$M_{k+1}(u) \leq M_k(u)$$

if $a_{k+1} < a_k$.

Now assume that the $u_j(t)$, $j = 1, 2, \ldots, k$, are defined already and belong to B_s if $0 \leq t < a_j(s_0 - s)$, $a_1 > a_2 > \ldots > a_k > 0$. Assume by induction, further, that

$$\|u_j(t) - u_{j-1}(t)\|_s \leq \varepsilon_j R \tag{3.31}$$

while $M_j(u_j - u_{j-1})$ is supposed to be finite.

In order to prove the existence of u_{k+1}, we choose any positive $a_{k+1} < a_k$. Later on we shall obtain a positive lower bound for $a_k - a_{k+1}$ restricting the possible choice of a_{k+1}. For an arbitrary s, $0 < s < s_0$, we define $\tilde{s}$ by

$$a_{k+1}(s_0 - s) = a_k(s_0 - \tilde{s}),$$

i.e., $s < \tilde{s} < s_0$. The k-th approximation $u_k(\tau)$ is defined and belongs to $B_{\tilde{s}}$ if $0 \leq \tau < a_k(s_0 - \tilde{s})$. In view of (3.31) and (3.28) we have

$$\|u_k(\tau) - u_0\|_{\tilde{s}} \leq \sum_{j=1}^{k} \|u_j(\tau) - u_{j-1}(\tau)\|_{\tilde{s}} \leq \sum_{j=1}^{k} \varepsilon_j R < R.$$

Consequently, $F(\cdot, u_k(\cdot))$ and $u_{k+1}(t)$ are defined and belong to B_s if $0 \leq t < a_k(s_0 - \tilde{s})$, i.e.,$0 \leq t < a_{k+1}(s_0 - s)$. The variable τ of

$F(\tau, u_k(\tau))$ varies in the interval $0 \leq \tau \leq t$. From (3.19) one gets, moreover, that

$$\|u_{k+1}(t) - u_k(t)\|_s \leq \int_0^t \|F(\tau, u_k(\tau)) - F(\tau, u_{k-1}(\tau))\|_s d\tau.$$

Now we intend to estimate the integrand of the last integral with the help of assumption (III) on $F(t,u)$ (cf. 3.2.). For this end we must take the norm of $u_k(\tau) - u_{k-1}(\tau)$ in a space with larger index. This index will be chosen in dependence on τ and will be denoted by $s(\tau)$. The numbers τ and $s(\tau)$ must be connected by

$$\tau < a_k(s_0 - s(\tau)).$$

This inequality may be rewritten as

$$s(\tau) < s_0 - \frac{\tau}{a_k}. \tag{3.32}$$

Notice that for any two real numbers c_1 and c_2 with $c_1 < c_2$ we have

$$c_1 + \frac{1}{2}(c_2 - c_1) < c_2.$$

Thus (3.32) as well as the inequality $s < s(\tau)$ are satisfied if we choose

$$s(\tau) = s + \frac{1}{2}(s_0 - \frac{\tau}{a_k} - s). \tag{3.33}$$

Applying (III) we get, finally, the estimate

$$\|u_{k+1}(t) - u_k(t)\|_s \leq C \int_0^t \frac{\|u_k(\tau) - u_{k-1}(\tau)\|_{s(\tau)}}{s(\tau) - s} d\tau \tag{3.34}$$

if $0 \leq t < a_{k+1}(s_0 - s)$. We supposed by induction that $M_k(u_k - u_{k-1})$ is finite. From the definition of M_k we get for $\tau > 0$ the estimate

$$\|u_k(\tau) - u_{k-1}(\tau)\|_{s(\tau)} \leq \frac{M_k(u_k - u_{k-1})}{\frac{a_k(s_0 - s(\tau))}{\tau} - 1}$$

Applying this inequality and substituting the expression (3.33) for $s(\tau)$, one obtains from (3.34) the inequality

$$\|u_{k+1}(t) - u_k(t)\|_s \leq 4Ca_kM_k(u_k - u_{k-1})t \int_0^t \frac{1}{(a_k(s_0 - s) - \tau)^2} d\tau, \tag{3.35}$$

where the factor τ in the numerator of the integrand has been estimated by t. Calculating this integral, we obtain

$$4a_kCM_k(u_k - u_{k-1})\left[\frac{1}{a_k(s_0-s)-t} - \frac{1}{a_k(s_0-s)}\right].$$

The expression in the brackets is equal to

$$\frac{t}{(a_k(s_0-s)-t)a_k(s_0-s)}.$$

Since t is restricted to the interval

$$0 \leq t < a_{k+1}(s_0 - s)$$

the last expression can be estimated by

$$\frac{a_{k+1}}{a_k} \frac{1}{a_k(s_0 - s) - t} .$$

Summarizing the last inequalities, we get for $0 < t < a_{k+1}(s_0 - s)$ the estimate

$$\| u_{k+1}(t) - u_k(t) \|_s \leq 4a_{k+1} \, C \, M_k(u_k - u_{k-1}) \frac{1}{\frac{a_k(s_0-s)}{t} - 1} . \tag{3.36}$$

Since $a_{k+1} < a_k$ we have

$$M_{k+1}(u_{k+1} - u_k)$$
$$= \sup_{0<s<s_0} \sup_{0<t<a_{k+1}(s_0-s)} \| u_{k+1}(t) - u_k(t) \|_s \left(\frac{a_{k+1}(s_0-s)}{t} - 1 \right)$$
$$\leq \sup_{0<s<s_0} \sup_{0<t<a_{k+1}(s_0-s)} \| u_{k+1}(t) - u_k(t) \|_s \left(\frac{a_k(s_0-s)}{t} - 1 \right) .$$

Thus the inequality (3.36) yields the estimate

$$M_{k+1}(u_{k+1} - u_k) \leq 4a_{k+1} \, C \, M_k(u_k - u_{k-1}) . \tag{3.37}$$

This estimate shows, in particular, that $M_{k+1}(u_{k+1} - u_k)$ is finite. Notice that (3.37) holds also with $j = k, k - 1, \ldots, 2$ instead of $j = k + 1$:

$$M_k(u_k - u_{k-1}) \leq 4a_k \, C \, M_{k-1}(u_{k-1} - u_{k-2}),$$
$$M_{k-1}(u_{k-1} - u_{k-2}) \leq 4a_{k-1} \, C \, M_{k-2}(u_{k-2} - u_{k-3}),$$
$$\vdots$$
$$M_2(u_2 - u_1) \leq 4a_2 \, C \, M_1(u_1 - u_0).$$

Substituting these inequalities into each other and taking into account the estimate (3.26) with $a = a_1$, one obtains

$$M_k(u_k - u_{k-1}) \leq (4C)^{k-1} a_k a_{k-1} \cdots a_1 \, K. \tag{3.38}$$

On the other hand, we have

$$\frac{a_k(s_0 - s)}{t} - 1 > \frac{a_k(s_0 - s)}{a_{k+1}(s_0-s)} - 1 = \frac{a_k - a_{k+1}}{a_{k+1}}$$

if $0 < t < a_{k+1}(s_0 - s)$. Therefore, (3.36) leads also to the estimate

$$\| u_{k+1}(t) - u_k(t) \|_s \leq 4 \frac{a_{k+1}^2}{a_k - a_{k+1}} \, C \, M_k(u_k - u_{k-1}) .$$

In view of (3.38) one obtains, finally,

$$\|u_{k+1}(t) - u_k(t)\|_s \leqq (4C)^k \frac{a_{k+1}^2 a_k \cdots a_1}{a_k - a_{k+1}} \cdot K.$$

This inequality yields a lower bound for $a_k - a_{k+1}$ guaranteeing the estimate

$$\|u_{k+1}(t) - u_k(t)\|_s \leqq \varepsilon_{k+1} R.$$

The last estimate is satisfied if

$$a_k - a_{k+1} \geqq \frac{1}{\varepsilon_{k+1}} (4C)^k \, a_{k+1}^2 a_k \cdots a_1 \, \frac{K}{R}. \tag{3.39}$$

Summarizing the above arguments, until now we have proved the following preliminary result:

Provided we can find a sequence of positive numbers $a_1, a_2, \ldots$ satisfying the condition (3.39), the functions $u_k = u_k(t)$ are defined by induction for each $k = 1, 2, \ldots$ and belong to B_s if only $0 \leqq t < a_k(s_0 - s)$.

In view of (3.31) and the triangle inequality in B_s we have for any pair l, k with $l > k$

$$\begin{aligned} &\|u_k(t) - u_l(t)\|_s \\ &\leqq \|u_k(t) - u_{k+1}(t)\|_s + \ldots + \|u_{l-1}(t) - u_l(t)\|_s \\ &\leqq (\varepsilon_{k+1} + \ldots + \varepsilon_l) R. \end{aligned} \tag{3.40}$$

Taking into account formula (3.28) we see, consequently, that the $u_k(t)$ form a fundamental sequence. The limit function

$$u = \lim_{k\to\infty} u_k$$

will be defined for $0 \leqq t < a_*(s_0 - s)$ and will belong to B_s, where

$$a_* = \lim_{k\to\infty} a_k.$$

The a_k have not also to satisfy the condition (3.39) (for each $k = 1, 2, \ldots$), but the limit function $u = u(t)$ has also to exist in an interval of positive length, i.e., a_* has to be positive. For this end we define a_{k+1} as the product

$$a_{k+1} = a_k(1 - \delta_k), \tag{3.41}$$

where δ_k satisfies the inequality $0 < \delta_k < 1$. Then a_* turns out to be the infinite product

$$a_* = a_1 \prod_{k=1}^{\infty} (1 - \delta_k).$$

It is well-known that such an infinite product converges and has a positive value if and only if

$$\sum_{k=1}^{\infty} \delta_k < +\infty . \qquad (3.42)$$

In view of the choice (3.41) the condition (3.39) may be rewritten in the relation

$$\delta_k \varepsilon_{k+1} \geq (4C)^k \, a_{k+1}^2 a_{k-1} a_{k-2} \cdots a_1 \frac{K}{R}, \quad k = 1, 2, \ldots,$$

connecting δ_k and ε_{k+1}. This relation is satisfied if

$$\delta_k \varepsilon_{k+1} \geq (4a_1C)^k a_1 \frac{K}{R} \qquad (3.43)$$

for each $k = 1, 2, \ldots$ According to (3.28) and (3.42) we have $\varepsilon_k \to 0$ and $\delta_k \to 0$ if $k \to +\infty$. Thus the left-hand side of the inequality (3.43) tends to zero, too, if k tends to $+\infty$. Consequently, the condition (3.43) can be satisfied for each k only if

$$4a_1C < 1.$$

Taking into account the conditions (3.23) and (3.29), where a must be replaced by a_1, we assume now that a_1 satisfies the inequalities

$$a_1 \leq \min\left(\frac{T}{s_0}, \frac{\varepsilon_1 R}{K}\right), \quad a_1 < \frac{1}{4C} . \qquad (3.44)$$

In order to complete the proof of the convergence of the u_k, we must find sequences $\varepsilon_1, \varepsilon_2, \ldots$ and $\delta_1, \delta_2, \ldots$ of positive numbers such that conditions (3.28), (3.42), and (3.43) are satisfied. In the next section we discuss some possibilities different from each other for satisfying these conditions.

3.4. Construction of sequences $\varepsilon_1, \varepsilon_2, \ldots$ and $\delta_1, \delta_2, \ldots$

In this section we construct convergent series

$$\sum_{k=1}^{\infty} \varepsilon_k \quad \text{and} \quad \sum_{k=1}^{\infty} \delta_k$$

consisting of positive summands, where the value of the first series does not exceed 1, whereas in the case of the second series we demand only that each δ_k is less than 1. Both series must be connected by the relations (3.43) for each $k = 1, 2, \ldots$ and, therefore, the summands ε_k, δ_k must be chosen big enough. The number a_1 entering into the condition (3.43) mus satisfy the condition (3.44).

First choice

Condition (3.28) is satisfied if we take

$$\varepsilon_k = \frac{1}{2^k} . \qquad (3.45)$$

Then (3.43) is satisfied, especially, if

$$\delta_k = 2(8a_1C)^k \, a_1 \frac{K}{R} \, . \tag{3.46}$$

In this case the condition (3.42) is satisfied if and only if

$$8a_1C < 1. \tag{3.47}$$

The last inequality is for a_1 more restrictive than the second inequality (3.44). In order to ensure that each δ_k is less than 1, we restrict a_1 by

$$2a_1 \frac{K}{R} \leq 1.$$

In view of $\varepsilon_1 = \frac{1}{2}$ this condition is satisfied if the first inequality (3.44) holds.

In view of (3.47) we get for a_1 the restrictions

$$a_1 \leq \min\left(\frac{T}{s_0}, \frac{R}{2K}\right), \quad a_1 < \frac{1}{8C} \tag{3.48}$$

instead of (3.44). Summarizing these estimates, we have proved the following result:

Provided a_1 satisfied condition (3.48) the numbers ε_k and δ_k may be defined by (3.45) and (3.46), respectively. Then the number a_* characterizing the existence interval of the limit function of the u_k is given by

$$a_* = a_1 \prod_{k=1}^{\infty} \left(1 - 2(8a_1C)^k \, a_1 \frac{K}{R}\right). \tag{3.49}$$

Second choice

Now we take

$$\varepsilon_1 = \frac{1}{2}, \qquad \delta_k = \lambda\varepsilon_{k+1}, \tag{3.50}$$

where λ is a certain positive constant. Then the condition (3.43) with the equality sign yields

$$\lambda\varepsilon_{k+1}^2 = (4a_1C)^k \, a_1 \frac{K}{R}, \qquad k = 1, 2, \ldots \tag{3.51}$$

The condition (3.42), moreover, is satisfied if only (3.28) is true. In view of formula (3.51) and the first equation (3.50), the condition (3.28) yields

$$\frac{1}{\sqrt{\lambda}} \sum_{k=1}^{\infty} \sqrt{a_1} \sqrt{\frac{K}{R}} \, (2\sqrt{a_1C})^k = \frac{1}{2}$$

if we take the equality sign in (3.28).

The second inequality (3.44) shows that the series in the last equation converges. Therefore, we get

$$\sqrt{\lambda} = 4a_1\sqrt{\frac{KC}{R}} \, \frac{1}{1 - 2\sqrt{a_1C}}$$

$$\varepsilon_{k+1} = \tfrac{1}{2}(2a_1C)^{k-1}\ (1 - 2\sqrt{a_1C}) \tag{3.52}$$

and

$$\delta_k = \frac{1}{2}\,\frac{K}{RC}\,\frac{(2\sqrt{a_1C})^{k+3}}{1-2\sqrt{a_1C}}\ , \tag{3.53}$$

where k = 1, 2, ... Each δ_k is less than 1 if only

$$\frac{1}{2}\,\frac{K}{RC}\,\frac{(2\sqrt{a_1C})^4}{1 - 2\sqrt{a_1C}} < 1,$$

i.e.,

$$2\sqrt{a_1C}\ \left(1 + \frac{1}{2}\,\frac{K}{RC}\ (2\sqrt{a_1C})^3\right) < 1.$$

In view of the second inequality (3.44) the last one is satisfied if we assume

$$2\sqrt{a_1C}\ \left(1 + \frac{1}{2}\,\frac{K}{RC}\right) < 1.$$

Thus we get the condition

$$4a_1C < \left(\frac{2RC}{2RC + K}\right)^2$$

instead of the second inequality (3.44). Summarizing these considerations, we obtain in the second case the condition

$$a_1 \leq \min\left(\frac{T}{s_0}, \frac{R}{2K}\right), \qquad a_1 < \frac{1}{4C}\left(\frac{2RC}{2RC + K}\right)^2. \tag{3.54}$$

Suppose condition (3.54) is satisfied. Take ε_k, δ_k in accordance with (3.52), (3.53) and the first equation (3.50). Then a_* is given by

$$a_* = a_1 \prod_{k=1}^{\infty} \left(1 - \frac{1}{2}\,\frac{K}{RC}\,\frac{(2\sqrt{a_1C})^{k+3}}{1 - 2\sqrt{a_1C}}\right). \tag{3.55}$$

Third choice

Take

$$\delta_k = \frac{\gamma}{(k+\nu)^{\mu}}\ , \tag{3.56}$$

where γ is a positive number and ν, μ are natural numbers. Provided μ is not smaller than 2, then condition (3.42) is automatically satisfied. Suppose further that

$$\gamma < (1 + \nu)^{\mu}.$$

Then each δ_k is less than 1. The condition (3.43) is satisfied, too, if we choose

$$\varepsilon_{k+1} = (4a_1C)^k\ a_1\,\frac{K}{R}\,\frac{1}{\gamma}\ (k + \nu)^{\mu}, \tag{3.57}$$

k = 1, 2, ... Then

$$\frac{\varepsilon_{k+2}}{\varepsilon_{k+1}} = 4a_1C\left(\frac{k + 1 + \nu}{k + \nu}\right)^{\mu} \leq 4a_1C\left(\frac{2 + \nu}{1 + \nu}\right)^{\mu}.$$

Now replace the second condition (3.44) by the stronger one

$$4a_1C\left(\frac{2+\nu}{1+\nu}\right)^{\mu} < 1. \tag{3.58}$$

This condition is sufficient for the convergence of the series (3.28). By virtue of (3.58) the condition (3.28) is satisfied if

$$\varepsilon_1 + 4a_1^2\frac{KC}{R}\frac{1}{\gamma}(1+\nu)^{\mu}\sum_{k=0}^{\infty}\left[4a_1C\left(\frac{2+\nu}{1+\nu}\right)^{\mu}\right]^k < 1.$$

Thus a_1 must satisfy the quadratic inequality

$$\sigma a_1^2 \leq (1-\varepsilon_1)(1-\tau a_1),$$

where

$$\sigma = 4\frac{KC}{R}\frac{1}{\gamma}(1+\nu)^{\mu},$$

$$\tau = 4C\left(\frac{2+\nu}{1+\nu}\right)^{\mu}.$$

The last inequality yields the estimate

$$a_1 \leq \frac{1-\varepsilon_1}{2}\frac{\tau}{\sigma}\left[\sqrt{1+\frac{4\sigma}{(1-\varepsilon_1)\tau^2}}-1\right]. \tag{3.59}$$

In view of (3.59) we get the following result:

Suppose that a_1 satisfies the inequalities

$$a_1 \leq \min\left[\frac{T}{s_0}, \frac{\varepsilon_1 R}{K}, \frac{1-\varepsilon_1}{2}\frac{\tau}{\sigma}\left(\sqrt{1+\frac{4\sigma}{(1-\varepsilon_1)\tau^2}}-1\right)\right],$$

$$a_1 < \frac{1}{4C}\left(\frac{2+\nu}{1+\nu}\right)^{\mu}, \tag{3.60}$$

where ε_1 is an arbitrary number, $0 < \varepsilon_1 < 1$. Then admissible $\varepsilon_2, \varepsilon_3, \ldots$ and $\delta_1, \delta_2, \ldots$ are given by (3.56) and (3.57), and the number a_* characterizing the convergence interval of the $u_k(t)$ is given by

$$a_* = a_1\prod_{k=1}^{\infty}\left[1-\frac{\gamma}{(k+\nu)^{\mu}}\right]. \tag{3.61}$$

3.5. Existence of solutions of initial value problems in scales of Banach spaces

Now assume that we have sequences $\varepsilon_1, \varepsilon_2, \ldots$ and $\delta_1, \delta_2, \ldots$ satisfying the conditions (3.28) and (3.42). Suppose further that a_1 is any positive number such that condition (3.43) is satisfied for each $k = 1, 2, \ldots$ Then we proved in section 3.3. that there exists a positive number a_* such that the $u_k = u_k(t)$ defined by (3.19) belong to B_s and converge in the interval

$$0 \leq t < a_*(s_0 - s).$$

In section 3.4. we proved that numbers a_1 as well as sequences $\varepsilon_1, \varepsilon_2, \ldots$

and $\delta_1, \delta_2, \dots$ with properties formulated above exist in any case. In three special cases for the choice of the sequences $\varepsilon_1, \varepsilon_2, \dots$ and $\delta_1, \delta_2, \dots$ regarded in section 3.4. the number a_1 may be chosen arbitrarily in some interval given by the sufficient conditions (3.48), (3.54), and (3.60), respectively. Then the positive number a_* characterizing the convergence interval is given by (3.49), (3.55), and (3.61), respectively.

In view of (3.39) the $u_k = u_k(t)$ converge uniformly in each B_s if $0 \leq t < a_*(s_0 - s)$. Carrying out the limiting process $l \to \infty$, the estimate (3.40) shows further that the norm of the difference of $u_k = u_k(t)$ and the limit function $u = u(t)$ can be estimated uniformly by

$$\|u_k(t) - u(t)\|_s \leq \sum_{j=k+1}^{\infty} \varepsilon_j \cdot R. \tag{3.62}$$

Now let s and t be given numbers connected by $0 \leq t < a_*(s_0 - s)$. Then there exists a number s' with $s < s' < s_0$ such that the relation $0 \leq t < a(s_0 - s')$ is satisfied, too. Applying (3.62) with s' instead of s and with τ instead of t, $0 \leq \tau \leq t$, we see that the $u_k = u_k(\tau)$ converge in $B_{s'}$, where the limit function is $u = u(\tau)$. On the other hand, in view of (III) we have

$$\|F(\tau, u_k(\tau)) - F(\tau, u(\tau))\|_s \leq \frac{C\|u_k(\tau) - u(\tau)\|_{s'}}{s' - s}.$$

Thus the $F(\tau, u_k(\tau))$ converge uniformly to $F(\tau, u(\tau))$ in B_s if $0 \leq \tau \leq t$. This convergence implies, finally, the convergence

$$\int_0^t d\tau \cdot F(\tau, u_k(\tau)) \to \int_0^t d\tau \cdot F(\tau, u(\tau))$$

for $k \to \infty$. The limiting process $k \to \infty$ in (3.19) leads, consequently, to the relation

$$u(t) = u_0 + \int_0^t d\tau \cdot F(\tau, u(\tau)),$$

i.e., we obtain the integral equation (3.17), which is equivalent to the statement that $u = u(t)$ is a solution of the initial value problem (3.2), (3.3).

That way we have proved the following theorem[1]):

Theorem. Suppose that the right-hand side $F(t,u)$ of the differential

---

[1]) Under the only assumptions (I), (II), (III) this theorem was proved by T. Nishida in his paper [47] which generalizes L. Nirenberg's one [45]. In order to characterize the convergence interval, in both papers only the special infinite product

$$\prod_{k=1}^{\infty}(1 - \delta_k) \quad \text{with} \quad \delta_k = \frac{1}{(k+1)^2}$$

is used, whereas the proof of the convergence of the successive approximations given in [65] makes use of more general infinite products.

equation (3.2) satisfies the conditions (I), (II), and (III) formulated in section 3.2. Then the limit function $u = u(t)$ of the functions $u_k = u_k(t)$ defined by (3.19) is a solution of the initial value problem (3.2), (3.3) belonging to B_s if

$$0 \leq t < a_*(s_0 - s),$$

where

$$a_* = a_1 \prod_{k=1}^{\infty} (1 - \delta_k).$$

3.6. Lower bounds for the length of the convergence interval

According to the theorem formulated in the previous section 3.5., the convergence interval of the successive approximations (3.19) is characterized by an infinite product. In order to get a lower bound for the length of this interval, we need estimates from below for the value of such products. If their factors are of the form

$$(1 - \alpha q^k),$$

where α and q are positive numbers satisfying the conditions

$$q < 1 \quad \text{and} \quad \alpha q < 1,$$

then the following estimate holds:

$$\prod_{k=1}^{\infty} (1 - \alpha q^k) > (1 - \alpha q)^{\frac{1}{1-q}}.$$

Proof. Using the power series expansion of the natural logarithm, one has

$$\ln \prod_{k=1}^{\infty} (1 - \alpha q^k) = \sum_{k=1}^{\infty} \ln(1 - \alpha q^k) = - \sum_{k=1}^{\infty} \sum_{j=1}^{\infty} \frac{1}{j}(\alpha q^k)^j.$$

Rearranging the double series on the right-hand side, we see that the last expression equals to

$$- \sum_{j=1}^{\infty} (\frac{1}{j}\alpha^j \sum_{k=1}^{\infty} q^{kj}) = - \sum_{j=1}^{\infty} \frac{1}{j}\alpha^j \frac{q^j}{1 - q^j} .$$

In view of

$$- \frac{1}{1 - q^j} > - \frac{1}{1 - q}$$

we get, finally, the inequality

$$\ln \prod_{k=1}^{\infty} (1 - \alpha q^k) > - \frac{1}{1 - q} \sum_{j=1}^{\infty} \frac{1}{j}\alpha^j q^j = \frac{1}{1 - q} \ln(1 - \alpha q)$$

from which our statement follows. It may be applied to the infinite products (3.49) and (3.55) which are obtained in the cases of the first and the second choice of the sequences $\varepsilon_1, \varepsilon_2, \ldots$ and $\delta_1, \delta_2, \ldots$ (cf.

section 3.4.). In order to estimate the infinite product (3.61) that occurs in the case of the third choice, we regard first the special case $\gamma = 1$, $\nu = 1$, $\mu = 2$. Denote the j-th partial product by p_j. Then we have

$$p_{j-2} = \frac{1}{2}\,\frac{j}{j-1}$$

since

$$p_{j-2}\left(1 - \frac{1}{j^2}\right) = \frac{1}{2}\,\frac{j}{j-1}\,\frac{j^2-1}{j^2} = \frac{1}{2}\,\frac{j+1}{j} = p_{j-1}.$$

Therefore, one obtains

$$\prod_{k=1}^{\infty}\left(1 - \frac{1}{(k+1)^2}\right) = \lim_{j\to\infty} p_j = \frac{1}{2}$$

and, consequently,

$$\prod_{k=1}^{\infty}\left(1 - \frac{1}{(k+1)^{\mu}}\right) \geq \frac{1}{2}$$

if $\mu \geq 2$. Estimates of the more general products

$$\prod_{k=1}^{\infty}\left(1 - \frac{\gamma}{(k+\nu)^2}\right),$$

where $0 < \gamma \leq (1+\nu)^2$ may be obtained from the last one by representing $k + \nu$ in the form

$$k + \nu = k'(1+\nu) + k''$$

with $0 \leq k'' \leq \nu$ because this representation implies

$$1 - \frac{\gamma}{(k+\nu)^2} > 1 - \frac{1}{k'^2} .$$

We must take into account, however, that all k with the same k' but with remainders k'' different from each other lead to the same factor $(1 - \frac{1}{k'^2})$. Consequently, the infinite product

$$\prod_{k'=2}^{\infty}\left(1 - \frac{1}{k'^2}\right) = \frac{1}{2}$$

appears $(\nu + 1)$ times in the estimate.

3.7. Uniqueness of the solution of initial value problems in scales of Banach spaces

Provided the right-hand side F(t, u) of the differential equation (3.2) satisfies the conditions (I), (II), and (III) formulated in section 3.2., we proved in 3.3. and 3.5. that the initial value problem (3.2), (3.3) is solvable. Now we are going to show that the solution is unique. For this end we assume that u, v are two solutions of the same initial value problem (3.2), (3.3) defined and belonging to B_s if $0 \leq t < a(s_0 - s)$. Suppose that the right-hand side F(t,u) satisfies at least condition

(III). Assume, further, that $F(t, u(t))$ and $F(t, v(t))$ depend continuously on t. Since the initial value problem (3.2), (3.3) is equivalent to the integral equation (3.17), the difference $u - v$ of the two given solutions satisfies the equation

$$u(t) - v(t) = \int_0^t d\tau\cdot(F(\tau, u(\tau)) - F(\tau, v(\tau))). \tag{3.63}$$

The last equation corresponds to that for the difference of two solutions of the same initial value problem in a fixed Banach space (cf. section 1.6.).

In order to prove the asserted uniqueness, we estimate $\|u(t) - v(t)\|_s$ by means of the representation (3.63) of the difference $u - v$. In this way we get

$$\|u(t) - v(t)\|_s \leq \int_0^t \|F(\tau, u(\tau)) - F(\tau, v(\tau))\|_s \, d\tau.$$

In view of assumption (III) we get further

$$\|u(t) - v(t)\|_s \leq C \int_0^t \frac{\|u(\tau) - v(\tau)\|_{s(\tau)}}{s(\tau) - s} \, d\tau, \tag{3.64}$$

where $s(\tau)$ is to choose suitably, i.e., $s < s(\tau) < s_0$. Now we want to estimate the integral (3.64) in a similar way as we estimated the integral in (3.34). For this end we have to estimate the norm $\|u(\tau) - v(\tau)\|_{s(\tau)}$ by $M(u - v)$, where the functional $M = M(u)$ had been defined in section 3.3. Immediately this is not possible here, however since we do not know whether $M(u - v)$ is finite or not. Diminishing the t-interval, however, an estimate of this type can be in fact obtained. For this end take any number s_0' with $0 < s_0' < s_0$. Then regard only those s for which $0 < s < s_0'$. In order to apply the estimate (3.64) with $s(\tau) = s_0'$, we must limit the variable τ to the interval

$$s_0' < s_0 - \frac{\tau}{a}$$

(cf. (3.32)). This inequality is satisfied for each τ with $0 \leq \tau \leq t$ if

$$s_0' < s_0 - \frac{t}{a}.$$

Therefore, we obtain for t the inequality

$$0 \leq t < a(s_0 - s_0'). \tag{3.65}$$

Now we intend to replace this inequality by a stronger one for which the right-hand side depends on s. It is clear that

$$a'(s_0' - s) < a(s_0 - s_0')$$

for each s with $0 < s < s_0'$ if

$$a's_0' < a(s_0 - s_0'). \tag{3.66}$$

Thus the inequality (3.65) is satisfied for all t with

$$0 \leq t < a'(s_0' - s), \tag{3.67}$$

where a' is a positive number satisfying the inequality

$$a' < \frac{a(s_0 - s_0')}{s_0'} . \tag{3.68}$$

In view of (3.66) the closed interval

$$0 \leq \tau \leq a's_0'$$

is a subinterval of the semi-open interval (3.65) in which u(t) - v(t) are elements of $B_{s_0'}$. Thus the supremum

$$S = \sup_{0 \leq \tau \leq a's_0'} \|u(\tau) - v(\tau)\|_{s_0'}$$

is finite[1]. Therefore formula (3.64) with $s(\tau) = s_0'$ leads to the estimate

$$\|u(t) - v(t)\|_s \leq \frac{CSt}{s_0' - s} \tag{3.69}$$

for each t belonging to the interval (3.67). Now define the modified functional M' = M'(w) (with respect to the interval (3.67)) by

$$M'(w) = \sup_{0<s<s_0'} \sup_{0<t<a'(s_0'-s)} \|w(t)\|_s \left(\frac{a'(s_0' - s)}{t} - 1\right). \tag{3.70}$$

By virtue of (3.69) the definition (3.70) with w = u - v implies

$$M'(u-v) \leq CS(a' - \frac{t}{s_0' - s}) \leq CSa'.$$

Hence M'(u-v) turns out to be finite. Therefore we may estimate the numerators $\|u(\tau) - v(\tau)\|_{s(\tau)}$ of the integrand in (3.64) by the modified functional. The quantity $s(\tau)$ in (3.64), however, must also be replaced by a modified $s'(\tau)$ which corresponds to the diminished interval (3.67). Analogously to (3.33) define

$$s'(\tau) = s + \frac{1}{2}(s_0' - \frac{\tau}{a'} - s). \tag{3.71}$$

Then in the diminished interval (3.67) we obtain for $\|u(t) - v(t)\|_s$ the estimate

$$\|u(t) - v(t)\|_s \leq C \int_0^t \frac{\|u(t) - v(t)\|_{s'(\tau)}}{s'(\tau) - s} d\tau \tag{3.72}$$

instead of (3.64). Taking into account the definition (3.70) of M', we get

---

[1]) If F(t,u) satisfies not only condition (III) but also condition (I) (cf. section 3.2.), and if we further suppose that $\|u(t) - u_0\|_s \leq R$, $\|v(t) - u_0\|_s \leq R$ for each s, where R is given by condition (I), then one has $\|u(t) - v(t)\|_s \leq 2R$. Hence it is superfluous to introduce the quantity S.

$$\|u(\tau) - v(\tau)\|_{s'(\tau)} \leq \frac{M'(u - v)}{\frac{a'(s_0' - s'(\tau))}{\tau} - 1}$$

Applying this inequality and substituting the definition (3.71) of $s'(\tau)$, one obtains from (3.72) the inequality

$$\|u(t) - v(t)\|_s \leq 4a'CM'(u - v)\ t \int_0^t \frac{d\tau}{(a'(s_0' - s) - \tau)^2}, \qquad (3.73)$$

where the factor τ in the numerator of the integrand is estimated by t.

We have got the last estimate in a similar way as we obtained the estimate (3.35) from (3.34) in section 3.3. (the quantities a_k and s_0 in section 3.3. are here replaced by a' and s_0'). Calculating the integral in (3.73), we get, finally, the inequality

$$\|u(t) - v(t)\|_s \leq 4a'CM'(u - v) \frac{t^2}{(a'(s_0' - s) - t)\ a'(s_0' - s)}.$$

Therefore, in view of (3.67) we obtain the inequality

$$\|u(t) - v(t)\|_s \leq 4a'CM'(u - v) \frac{1}{\frac{a'(s_0' - s)}{t} - 1}$$

that is analogous to (3.36) (the corresponding calculations in the case of (3.36) are carried out in section 3.3.). From the last estimate one obtains further

$$\|u(t) - v(t)\|_s \left(\frac{a'(s_0' - s)}{t} - 1\right) \leq 4a'CM'(u - v)$$

for any s, t with $0 < s < s_0'$, $0 \leq t < a'(s_0' - s)$. Once more taking into consideration the definition (3.70) of the functional M', this inequality yields

$$M'(u - v) \leq 4a'CM'(u - v).$$

Provided

$$4a'C < 1 \qquad (3.74)$$

the last inequality proves

$$M'(u - v) = 0.$$

Thus we have proved the following result: The differences u(t) - v(t) interpreted as elements of B_s, $0 < s < s_0'$, vanish in the interval (3.67) having the length $a'(s_0' - s)$, where the number a' is chosen sufficiently small.

Now let t, s be any fixed pair for which u(t), v(t) are defined and belong to B_s, i.e., t belongs to the interval

$$0 \leq t < a(s_0 - s) \qquad (3.75)$$

with the length $a(s_0 - s)$. Now choose any s_0' with $s < s_0' < s_0$. In dependence on s_0' choose further a number a' satisfying the inequalities (3.68) and (3.74). In accordance with the result concerning the identical vanishing of the difference of two solutions we know that $u(t) - v(t)$ vanishes identically in an interval of length $a'(s_0' - s)$ if the initial value is equal to zero. Subdividing the whole given interval (3.75) into subintervals the lengths of which are less than $a'(s_0' - s)$ with suitably chosen s_0' and a', we conclude that $u(t) - v(t)$ must vanish identically in the whole interval (3.75). Thus we have proved the following theorem:

Theorem. Two solutions $u = u(t)$ and $v = v(t)$ of the initial value problem (3.2), (3.3) belonging to B_s in the interval (3.75) must coincide.

Remark 1. The theorem shows that the initial value problem (3.2), (3.3) is uniquely solvable in a given scale of Banach spaces. On the other hand the theorem does not exclude that there could exist other solutions not belonging to the scale, possibly (cf. also 10.3.).

Remark 2. The uniqueness can also be proved by applying A. Crodel's generalization of the Gronwall lemma (cf. 10.4.).

3.8. Linear differential equations in scales of Banach spaces

In the sections 3.3. and 3.5. we have proved that the successive approximations (3.19) converge to the solution of the initial value problem (3.2), (3.3). In accordance with condition (I) of section 3.2. we assumed that the right-hand side $F(t,u)$ is defined only for $u \in B_s$ satisfying the inequality $\|u - u_0\|_s \leq R$. In order to ensure the existence of all approximations $u_k = u_k(t)$, we had to diminish the t-interval step by step. In the following we show that such diminutions are not necessary if the right-hand side $F(t,u)$ is linear in u and is defined for all u belonging to B_s. For this end we replace condition (I) of 3.2. by the following one:

(I') There exists a positive number T such that for any pair s', s with $0 < s' < s < s_0$ the right-hand side $F(t,u)$ defines a continuous mapping of

$$\{t : 0 \leq t \leq T\} \times B_s$$

into $B_{s'}$.

This condition implies that all approximations (3.18), (3.19) are defined and belong to B_s for all t with $0 \leq t \leq T$. This can be proved by induction in the following way: In order to prove that $u_{k+1}(t)$ belongs to B_s, we choose $\tilde{s}$ with $s < \tilde{s} < s_0$. Since $u_k(t)$ belongs to $B_{\tilde{s}}$ we see from (I') that $F(\tau, u_k(\tau))$ and, consequently, $u_{k+1}(t)$ belong to B_s.

A certain diminution of the t-interval will be necessary only in order to ensure the convergence of the u_k. Applying condition (II), from (3.18) it follows immediately that

$$\|u_1(t) - u_0\|_s \leqq \frac{Kt}{s_0 - s} \tag{3.76}$$

(cf. (3.2)). Now we are going to prove by induction that

$$\|u_{k+1}(t) - u_k(t)\|_s \leqq \frac{K}{Ce}\left(\frac{Cet}{s_0 - s}\right)^{k+1} \tag{3.77}$$

for each k = 0, 1, 2, ..., where C is the constant entering into condition (III). In view of (3.76) the asserted estimate (3.77) is true for k = 0. Assume now that (3.77) holds for a fixed k, for each t with $0 \leqq t \leqq T$ and for each s, $0 < s < s_0$. First the definition (3.19) and the corresponding equation with k + 1 instead of k lead to the representation

$$u_{k+2}(t) - u_{k+1}(t) = \int_0^t d\tau \cdot \left(F(\tau, u_{k+1}(\tau)) - F(\tau, u_k(\tau))\right)$$

and, therefore, to the estimate

$$\|u_{k+2}(t) - u_{k+1}(t)\|_s$$

$$\leqq \int_0^t \|F(\tau, u_{k+1}(\tau)) - F(\tau, u_k(\tau))\|_s d\tau.$$

Taking into account condition (III), one obtains, moreover, the estimate

$$\|u_{k+2}(t) - u_{k+1}(t)\|_s$$

$$\leqq \frac{C}{\tilde{s} - s} \int_0^t \|u_{k+1}(\tau) - u_k(\tau)\|_{\tilde{s}} d\tau,$$

where $\tilde{s}$ is any number satisfying the inequality $s < \tilde{s} < s_0$. Using the estimate (3.77) that holds by assumption with $\tilde{s}$ instead of s, we get

$$\|u_{k+2}(t) - u_{k+1}(t)\|_s$$

$$\leqq \frac{C}{\tilde{s} - s} K C^k e^k \frac{1}{(s_0 - s)^{k+1}} \frac{t^{k+2}}{k+2} .$$

Since

$$\left(\frac{k + 2}{k + 1}\right)^{k+1} = \left(1 + \frac{1}{k + 1}\right)^{k+1}$$

the choice

$$\tilde{s} = s + \frac{1}{k + 2}(s_0 - s)$$

yields the estimate

$$\|u_{k+2}(t) - u_{k+1}(t)\|_s \leqq K C^{k+1} e^{k+1}\left(\frac{t}{s_0 - s}\right)^{k+2},$$

i.e., (3.77) holds for each k.

Provided that $0 \leq t \leq T$ we now restrict the variable t to the semi-open interval

$$0 \leq t < \frac{1}{Ce}(s_0 - s). \tag{3.78}$$

Then the quotient

$$\frac{Cet}{s_0 - s}$$

is less than 1 and, consequently, the series

$$\sum_{k=0}^{\infty} \left(\frac{Cet}{s_0 - s}\right)^{k+1}$$

converges, and the convergence is uniform in each closed subinterval of (3.78). By virtue of (3.77) the $u_k(t)$ form a convergent sequence in the whole interval (3.78). As we proved in section 3.5., the limit function turns out to be a solution of the initial value problem (3.2), (3.3).

Notice that condition (I') is more restrictive than condition (I), i.e., (I) is satisfied with each $R > 0$ if (I') is satisfied. Hence the uniqueness proof carried out in the previous section 3.7. is applicable if the right-hand side F(t,u) satisfies the conditions (I'), (II), and (III). Summarizing these arguments, the following theorem has been proved[1]):

Theorem. The initial value problem (3.3) for a linear differential equation (3.2) is uniquely solvable in a scale of Banach spaces B_s, $0 < s < s_0$, if the right-hand side F(t,u) satisfies the condition (I'), (II), and (III). The solution may be obtained by successive approximations and belongs to B_s if $0 \leq t \leq T$ and $0 \leq t < \frac{1}{Ce}(s_0 - s)$.

We would like to emphasize that the condition (I') as well as the corresponding condition (I) in the theorem in section 3.5. are necessary mainly in order to ensure that F(t, u(t)) depends condinuously on t if $u = u(t)$ is continuous. In special cases these conditions may be replaced by weaker ones. In section 7.4., for instance, we inverstigate differential operators with coefficients depending only continuously on t, whereas the desired solutions are Hölder-continuous for fixed t. In this case the method of successive approximations is applicable, too, although condition (I') is not necessarily satisfied.

[1]) The proof follows that given in F. Treves' book [61]. The basic idea of this proof is to derive the estimate (3.77). - Let us furthermore remark that nowhere we need the assumption that F(t,u) is linear in u. Right-hand sides F(t,u), however, not being linear in u but satisfying the conditions (I'), (II), and (III) do not exist, probably.

4. THE CLASSICAL CAUCHY-KOVALEVSKAYA THEOREM

4.1. Statement of the problem

The classical Cauchy-Kovalevskaya theorem deals with complex-valued functions

$$w_j = w_j(t, z_1, \ldots, z_n), \quad j = 1, \ldots, m,$$

depending on one real variable t and n complex variables $z_1, \ldots, z_n$. Assume that the functions $w_j = w_j(t, z_1, \ldots, z_n)$ looked for are solutions of differential equations of the following type:

The k-th derivatives

$$\frac{\partial^k w_j}{\partial t^k}, \; j = 1, \ldots, m, \tag{4.1}$$

can be expressed by given complex-valued functions f_j that may depend not only on $t, z_1, \ldots, z_n, w_1, \ldots, w_m$ but also on the derivatives of the unknown functions up to the order k with the exception of the derivatives (4.1), i.e.,

$$\frac{\partial^k w_j}{\partial t^k} = f_j(t, z_1, \ldots, z_n, w_1, \ldots, w_m, \frac{\partial w_1}{\partial t}, \ldots, \frac{\partial^k w_m}{\partial z_n^k}), \tag{4.2}$$

$j = 1, \ldots, m$. We suppose that the w_j depend holomorphically on the z_i so that the partial complex derivatives with respect to z_i may be understood as ordinary complex derivatives in the z_i-plane. For short we denote the vector $(z_1, \ldots, z_n)$ by z and the vector $(w_1, \ldots, w_m)$ by w. Let p further be the vector consisting of the lexicographically ordered derivatives of the $w_1, \ldots, w_m$,

$$p = \left(\frac{\partial w_1}{\partial t}, \frac{\partial w_1}{\partial z_1}, \ldots, \frac{\partial^k w_m}{\partial z_n^k}\right), \tag{4.3}$$

where the derivatives (4.1) are omitted. Then the system (4.2) of differential equations may be rewritten as

$$\frac{\partial^k w}{\partial t^k} = f(t, z, w, p), \tag{4.4}$$

where the vector $(f_1, \ldots, f_m)$ is denoted by f.

The classical Cauchy-Kovalevskaya theorem formulates sufficient conditions under which the following initial value problem is solvable:

We look for a solution of (4.2) for which the values w(t,z) of the solution itself and those of its first k - 1 derivatives with respect to the variable t are prescribed at t = 0, i.e.,

$$w(0,z) = \phi^{(0)}(z),$$

$$\frac{\partial w}{\partial t}(0,z) = \phi^{(1)},$$
$$\vdots \qquad (4.5)$$
$$\frac{\partial^{k-1} w}{\partial t^{k-1}}(0,z) = \phi^{(k-1)}(z),$$

where $\phi^{(o)}, \ldots, \phi^{(k-1)}$ are given vectors.

In order to ensure the solvability of the initial value problem (4.4), (4.5), now we are going to formulate sufficient conditions on the right-hand side f and the initial vectors $\phi^{(i)}$, $i = 0, 1, \ldots, k-1$. Suppose that t varies in a closed interval, $0 \leq t \leq T$. Suppose further that z varies in a domain G_z of the n-dimensional complex space $\mathbb{C}^n$, whereas w varies in the domain G_w of the $\mathbb{C}^m$.

First we would like to calculate the number of the components of the vector (4.3). Notice that the binomial coefficients $\binom{\sigma}{\tau}$ satisfy the following relation

$$\sum_{i=1}^{k} \binom{q+i}{i} = \binom{q+k+1}{k} - 1 . \qquad (4.6)$$

One sees immediately that this relation is true for $k = 1$. Supposing it holds for k, in view of the well-known formula

$$\binom{\sigma}{\tau} + \binom{\sigma}{\tau+1} = \binom{\sigma+1}{\tau+1}$$

one obtains

$$\sum_{i=1}^{k+1} \binom{q+i}{i} = \binom{q+k+1}{k} - 1 + \binom{q+k+1}{k+1} = \binom{q+k+2}{k+1} - 1.$$

Hence (4.6) has been proved by induction for each k. On the other hand, a continuously differentiable function depending on σ variables possesses

$$\binom{\sigma+i-1}{i} \qquad (4.7)$$

different derivatives of the order i. It may be added that this quantity (4.7) is the number of combinations of σ things, i at a time, when repetitions are allowed [1]).

In view of (4.7) there are

$$m\binom{n+i}{i}$$

---

[1]) Denote this number by $\tilde{C}_i^{\sigma}$. Then we have, obviously,

$$\tilde{C}_{i+1}^{\sigma} = \tilde{C}_i^{\sigma} + \tilde{C}_i^{\sigma-1} + \ldots + \tilde{C}_i^{1}.$$

Applying the relation (4.6), one may prove by induction with respect to i that $\tilde{C}_i^{\sigma}$ is equal to (4.7).

derivatives of the order i, i = 1, ..., k - 1, entering into the vector (4.3), whereas only

$$m\left[\binom{n+k}{k} - 1\right]$$

derivatives of order k enter into (4.3) because the derivatives (4.1) are missing. Using the relation (4.6) one obtains, finally, that the vector (4.3) possesses

$$d = m\left[\binom{n+k+1}{k} - 2\right]$$

components. Thus p varies in the complex space $\mathbb{C}^d$. Suppose in the following that the right-hand side f(t,z,w,p) possesses the following properties:

a) f(t,z,w,p) is defined and continuous in
   $\{t : 0 \leq t \leq T\} \times G_z \times G_w \times G_p$, where G_p is a domain in $\mathbb{C}^d$.

b) f(t,z,w,p) depends holomorphically on all components of z, w, and p.

c) f(t,z,w,p) and its first order derivatives with respect to all components of z, w, and p are uniformly bounded and satisfy a uniform Lipschitz condition with respect to all components of w and p.

Now we are going to formulate conditions about the initial vectors $\phi^{(i)}$, i = 0, 1, ..., k - 1. For this end we need the vector p (cf. (4.3)) at the point t = 0. In view of the initial conditions (4.5) all derivatives of the w_j, j = 1, ..., m, at the point t = 0 may be replaced by $\phi_j^{(i)}$ and their derivatives, where $\phi_1^{(i)}, \ldots, \phi_m^{(i)}$ are the components of $\phi^{(i)}$. E. g., we have

$$\frac{\partial^2 w_j}{\partial t^2}(0,z) = \phi_j^{(2)}(z)$$

and

$$\frac{\partial^3 w_j}{\partial t \partial z_1 \partial z_2}(0,z) = \frac{\partial^2 \phi_j^{(1)}}{\partial z_1 \partial z_2}(z).$$

Let $p^{(o)}$ be the vector which we obtain by these substitutions, i.e.,

$$p^{(o)}(z) = \left[\phi_1^{(1)}(z), \ldots, \frac{\partial^k \phi_m^{(o)}}{\partial z_n^k}(z)\right].$$

Suppose that the initial vectors $\phi^{(o)}, \ldots, \phi^{(k-1)}$ possess the following properties:

a) They are holomorphic vectors defined in G_z.

b) For each $z \in G_z$ the values $\phi^{(0)}(z)$ belong to a compact subset of G_w.

c) Analogously, the values $p^{(0)}(z)$ belong to a compact subset of G_p if $z \in G_z$.

We remark that in view of the holomorphy of the $\phi_j^{(i)}$ the partial derivatives with respect to $z_1, \ldots, z_n$ may be interpreted as ordinary complex derivatives. Note that we need the assumptions b) and c) about the initial vectors in order to ensure that the right-hand side f of the differential equation (4.4) is defined in a neighbourhood of the initial values and their derivatives up to the order k - 1.

Finally we assume that together with the domain G_z there is given a family of subdomains $G_{z,s}$, $0 < s < s_0$, satisfying the three conditions a), b), c) on families of subdomains formulated in section 2.2. If, for instance, G_z is the polydisk

$$G_z = \{z : |z_i| < r_i\},$$

then a family of subdomains possessing the desired properties is given by

$$G_{z,s} = \{z : |z_i| < sr_i\}, \qquad 0 < s < 1.$$

Indeed, if z' belongs to $G_{z,s'}$ and z belongs to the boundary of $G_{z,s}$, where $s' < s$, then one obtains

$$|z_i - z_i'| \geqq (s - s')r_i$$

and, consequently,

$$d(G_{z,s'}, \partial G_{z,s}) \geqq \left(\sum_{i=1}^{n} r_i^2\right)^{\frac{1}{2}}(s - s').$$

This is, however, the estimate (2.5). Thus the condition b) on the family of subdomains is satisfied. The remaining conditions a) and c) are obvious.

Provided the right-hand side f(t,z,w,p) and the initial functions $\phi^{(0)}$, ..., $\phi^{(k-1)}$ satisfy the assumptions formulated above, the following theorem (classical Cauchy-Kovalevskaya theorem) holds:

<u>Theorem.</u> The initial value problem (4.4), (4.5) is solvable by a vector $w = w(t,z)$ defined and holomorphic in $G_{z,s}$ if t belongs to the interval

$$0 \leqq t < a(s_0 - s),$$

where a is a positive number depending on both the right-hand side of the differential equation and the initial functions.

4.2. Reduction of Cauchy-Kovalevskaya systems to quasilinear first order systems

In order to prepare the proof of the theorem formulated in the preceding section we simplify the initial value problem (4.4), (4.5) in three steps.

First step: Reduction to the initial values zero

Introduce a vector $\tilde{w}(t,z)$ instead of $w(t,z)$ by

$$w(t,z) = \tilde{w}(t,z) + \sum_{i=0}^{k} \frac{1}{i!} t^i \phi^{(i)}(z). \tag{4.8}$$

Then the initial condition (4.5) is equivalent to

$$\begin{aligned} \tilde{w}(0,z) &= 0, \\ \frac{\partial \tilde{w}}{\partial t}(0,z) &= 0, \\ &\vdots \\ \frac{\partial^{k-1}\tilde{w}}{\partial t^{k-1}}(0,z) &= 0 \end{aligned}$$

and vice versa. The substitution (4.8) changes also the differential equation (4.4). Nevertheless, the additional assumptions on $f(t,z,w,p)$ formulated in the previous section 4.1. remain true after the substitution. Therefore without any loss of generality we may assume that the initial functions are identically equal to zero.

Second step: Reduction to first order systems

In order to prove that the system (4.4) may be reduced to a first order system, it is sufficient to show that the order of (4.4) can be reduced by 1. For this end in addition to the vector w we introduce the vectors

$$w^{(i)} = \frac{\partial w}{\partial z_i}, \quad i = 1, \ldots, n, \tag{4.9}$$

and

$$w^{(n+1)} = \frac{\partial w}{\partial t}. \tag{4.10}$$

From these definitions we get $(k \geq 2)$

$$\frac{\partial^{k-1} w}{\partial t^{k-1}} = \frac{\partial^{k-2} w^{(n+1)}}{\partial t^{k-2}},$$

$$\frac{\partial^{k-1} w^{(i)}}{\partial t^{k-1}} = \frac{\partial^{k} w}{\partial t^{k-1}\partial z_i} = \frac{\partial^{k-1} w^{(n+1)}}{\partial t^{k-2}\partial z_i},$$

$$\frac{\partial^{k-1} w^{(n+1)}}{\partial t^{k-1}} = \frac{\partial^{k} w}{\partial t^{k}}.$$

Hence it follows that the original system (4.4) may be rewritten as

$$\frac{\partial^{k-1}w}{\partial t^{k-1}} = \frac{\partial^{k-2}w^{(n+1)}}{\partial t^{k-2}},$$

$$\frac{\partial^{k-1}w^{(i)}}{\partial t^{k-1}} = \frac{\partial^{k-1}w^{(n+1)}}{\partial t^{k-2}\,\partial z_i}, \quad i = 1, \ldots, n, \tag{4.11}$$

$$\frac{\partial^{k-1}w^{(n+1)}}{\partial t^{k-1}} = f(t,z,w,p).$$

In order to reproduce the form (4.3) of the vector p in the case of the system (4.11), we must replace the derivatives of the order k by derivatives of the order k - 1,where never a derivative occurs with k - 1 differentiations with respect to t. Such replacement is in fact possible. To this end we regard, firstly, a derivative of order k that contains at least one differentiation with respect to t. This derivative contains at least one differentiation with respect to one of the variables $z_1, \ldots, z_n$ because the same is true for each derivative of the order k in the vector (4.3). Using the definition (4.10) we consequently obtain a derivative of the order k - 1 that contains at most k - 2 differentiations with respect to t. Secondly we regard a derivative of the order k not containing differentiations with respect to t. Using now the definition (4.9), we obtain a derivative of the order k - 1 without any differentiations with respect to t. The derivative $\frac{\partial^{k-1}w}{\partial t^{k-1}}$, finally, may be replaced by $\frac{\partial^{k-2}w^{(n+1)}}{\partial t^{k-2}}$.

Summarizing these substitutions, we see that the original system (4.4) of the order k may be reduced to the system (4.11) of the order k - 1 possessing analogous properties.

By virtue of the first step we may assume (without any loss of generality) that the vector w = w(t,z) satisfies the homogeneous initial condition

$$\begin{aligned} w(0,z) &= 0, \\ \frac{\partial w}{\partial t}(0,z) &= 0, \\ &\vdots \\ \frac{\partial^{k-1}w}{\partial t^{k-1}}(0,z) &= 0. \end{aligned} \tag{4.12}$$

Then in view of the definition (4.9) and (4.10) the vectors $w^{(i)}$, i = 1, ..., n, n + 1, satisfy homogeneous initial conditions

$$\begin{aligned} w^{(i)}(0,z) &= 0, \\ &\vdots \end{aligned}$$

(4.13)

$$\frac{\partial^{k-2} w^{(i)}}{\partial t^{k-2}}(0,z) = 0,$$

too.

Conversely assume that $w, w^{(1)}, \ldots, w^{(n)}, w^{(n+1)}$ is a solution to the system (4.11), where the initial values vanish identically, i.e., w satisfies the k - 1 first equations (4.12), whereas the vectors $w^{(1)}, \ldots, w^{(n+1)}$ satisfy the conditions (4.13). Since in view of (4.10) the (k-1)-th derivative of w with respect to t equals to the (k-2)-th derivative of $w^{(n+1)}$, the vector w satisfies also the last equation (4.12). It remains to prove that w is a solution to the original system (4.2) if $w, w^{(1)}, \ldots, w^{(n+1)}$ is a solution of the (k-1)-th order system (4.11). This follows immediately from the last equation (4.11) if one is able to show that the system (4.11) implies the equations (4.9) and (4.10) which allow us to express the components of the vector p in the last equations (4.11) by the derivatives of w alone.

In order to deduce equation (4.10) from the system (4.11), we rewrite the first equation of the system (4.11) in the form

$$\frac{\partial^{k-2}}{\partial t^{k-2}}\left(\frac{\partial w}{\partial t} - w^{(n+1)}\right) = 0.$$

Thus for fixed z the components of the vector $\frac{\partial w}{\partial t} - w^{(n+1)}$ are polynomials in t whose degree is at most equal to k - 3. On the other hand, the initial values of the vector in question and those of the derivatives with respect to t (even up to the order k - 2) vanish, too. Hence (4.10) holds.

Differentiating the first equation (4.11) with respect to z_i and subtracting the second equation (4.11), we get, analogously,

$$\frac{\partial^{k-1}}{\partial t^{k-1}}\left(\frac{\partial w}{\partial z_i} - w^{(i)}\right) = 0.$$

Hence $\frac{\partial w}{\partial z_i} - w^{(i)}$ turns out to be a polynomial in t (of a degree not larger than k - 2) whose initial values vanish. Thus also the equations (4.9) are true.

Therefore the k-th order system (4.2) and the (k-1)-th order system (4.11) are equivalent.

Notice that there are different possibilities to express the derivatives of w by the functions $w^{(i)}, w^{(n+1)}$ defined by (4.9) and (4.10). We have, for instance,

$$\frac{\partial^2 w}{\partial t \partial z_1} = \frac{\partial w^{(1)}}{\partial t}$$

as well as

$$\frac{\partial^2 w}{\partial t \partial z_1} = \frac{\partial w^{(n+1)}}{\partial z_1}.$$

Thus the system (4.11) is not uniquely determined by the original one (4.2). Nevertheless the systems (4.2) and (4.11) are equivalent if they are reduced into each other by the same substitutions.

Third step: Reduction to a quasilinear first order system

In view of the preceding step we may restrict ourselves to the first order system

$$\frac{\partial w}{\partial t} = f(t,z,w,\frac{\partial w}{\partial z_1},\dots,\frac{\partial w}{\partial z_n}) \tag{4.14}$$

instead of to the system (4.4) of the order k. The right-hand side does not depend linearly on the derivatives $\frac{\partial w}{\partial z_i}$, in general. In order to reduce this system to a quasilinear one, we interpret the n derivatives $\frac{\partial w}{\partial z_i}$, i = 1, ..., n, as unknown vectors $w^{(i)}$ (cf. formula (4.9)). Hence equation (4.14) is turned into

$$\frac{\partial w}{\partial t} = f(t,z,w,w^{(1)},\dots,w^{(n)}).$$

Writing this equation componentwise, we get

$$\frac{\partial w_k}{\partial t} = f_k(t,z,w_1,\dots,w_m,w_1^{(1)},\dots,w_m^{(n)}), \tag{4.15}$$

k = 1, ..., m, where the components of w and $w^{(i)}$ are denoted by w_1, ..., w_m and $w_1^{(i)}$, ..., $w_m^{(i)}$, respectively. In order to preserve the form of the differential equations we must add equations for the derivatives $\frac{\partial w_k^{(i)}}{\partial t}$ to the system (4.15). We get these equations by differentiating the equations (4.15) with respect to z_i. That way by using the chain rule one obtains

$$\frac{\partial w_k^{(i)}}{\partial t} = \frac{\partial f_k}{\partial z_i}(\dots) + \sum_{j=1}^{m} \frac{\partial f_k}{\partial w_j}(\dots)\frac{\partial w_j}{\partial z_i} + \sum_{\nu=1}^{n}\sum_{j=1}^{m} \frac{\partial f_k}{\partial w_j^{(\nu)}}(\dots)\frac{\partial w_j^{(\nu)}}{\partial z_i}, \tag{4.16}$$

i = 1, ..., n, k = 1, ..., m, where the right-hand side depends on the same variables as the right-hand side of (4.15) does. Provided the initial values of the solution w = w(t,z) of the differential equation (4.14) vanish identically, the same is true for the initial values of the vectors $w^{(i)}$ introduced additionally, i.e.,

$$w^{(i)}(0,z) = 0.$$

Now we are going to prove, conversely, that w is a solution to (4.14) if the components of w, $w^{(1)}$, ..., $w^{(n)}$ are a solution of the enlarged system (4.15), (4.16) satisfying homogeneous initial conditions. Taking into account the way of obtaining the system (4.16), it is obvious that

$$\frac{\partial}{\partial t}\left(\frac{\partial w_k}{\partial z_i} - w_k^{(i)}\right)$$

vanishes identically. On the other hand all initial values of $\frac{\partial w_k}{\partial z_i} - w_k^{(i)}$ are equal to zero at the point $t = 0$, too. Therefore, the differences $\frac{\partial w_k}{\partial z_i} - w_k^{(i)}$ vanish everywhere. Hence the equations (4.15) coincide with (4.14). Consequently, the system (4.15), (4.16) turns out to be equivalent to equation (4.14) (in the case of homogeneous initial conditions).

Summarizing the above arguments, we have obtained that the differential equations (4.2) may be replaced by the quasilinear first order equations

$$\frac{\partial w_j}{\partial t} = \sum_{i=1}^{n} \sum_{k=1}^{m} a_{ik}^{(j)}(t,z,w)\frac{\partial w_k}{\partial z_i} + a_0^{(j)}(t,z,w), \tag{4.17}$$

$j = 1, \ldots, m$. Provided the right-hand sides of (4.2) satisfy the conditions formulated in section 4.1., the coefficients $a_{ik}^{(j)}(t,z,w)$, $a_0^{(j)}(t,z,w)$ possess the following properties:

a) They are defined and continuous in

$$\{t : 0 \leq t \leq T\} \times G_z \times G_w.$$

b) They depend holomorphically on all components of z and w.

c) They are uniformly bounded and satisfy a uniform Lipschitz condition with respect to all components of w.

Since the initial values are reduced to zero, the assumptions on the initial values formulated in section 4.1. are simplified to the only condition that $(0, \ldots, 0)$ is an interior point of G_w, i.e., that there exists a positive number R such that the closed polydisk

$$\{w = (w_1, \ldots, w_m) : |w_j| \leq R\} \tag{4.18}$$

is contained in the domain G_w in which the variable w varies.

4.3. Proof of the classical Cauchy-Kovalevskaya theorem

Now we intend to prove the classical Cauchy-Kovalevskaya theorem formulated in section 4.1. In view of the previous section 4.2. it is sufficient to carry out the proof in the case of the quasilinear first order system (4.17) with identically vanishing initial values. Suppose that the coefficients $a_{ik}^{(j)}$, $a_0^{(j)}$ satisfy the conditions formulated at the end of the preceding section. Let A be an upper bound for the modules of all coefficients $a_{ik}^{(j)}$, $a_0^{(j)}$ entering into the differential equations (4.17). Let, moreover, l be a common Lipschitz constant of all coefficients with respect to $w_1, \ldots, w_m$ existing by assumption. Suppose further that $G_{z,s}$, $0 < s < s_0$, is a family of subdomains of G_z mentioned above (cf. sections 2.2. and 4.1.).

The announced proof will be based on the general theorem on the solution of initial value problems in scales of Banach spaces (cf. section 3.5.). Thus we have to introduce a suitable scale of Banach spaces in which the initial value problem

$$w(0,z) = \phi_0(z) = 0 \tag{4.19}$$

for the system (4.17) can be solved by successive approximations. Since we look for vectors $w = (w_1, \ldots, w_m)$ depending holomorphically on z we introduce the space H_s of all vectors w holomorphic in $G_{z,s}$ and continuous in $\overline{G}_{z,s}$. The space H_s equipped with the supremum norm (maximum norm)

$$\|w\|_s = \max_{1\leq j\leq m} \sup_{G_{z,s}} |w_j(z)|$$

turns out to be a Banach space. In the case of one complex variable this statement was proved in section 2.1. The same statement is true, however, in the case of holomorphic vectors depending on several complex variables $z_1, \ldots, z_n$. Analogously as in the case of one complex variable, the proof is based on the Cauchy integral formula that also holds for holomorphic functions of several complex variables.

First we are going to deduce <u>Cauchy's integral formula in the multidimensional complex analysis</u>. For this end we assume that $w_k = w_k(z_1,\ldots,z_n)$ is holomorphic in $G_{z,s}$. Let $z_o = (z_{o1},\ldots,z_{on})$ be a fixed point of $G_{z,s}$. Regard the polydisk

$$\{z = (z_1,\ldots,z_n) : |z_j - z_{oj}| \leq r_j\}. \tag{4.20}$$

The whole closed polydisk (4.20) belongs to $G_{z,s}$ if the radii are sufficiently small. Take any interior point $z' = (z_1',\ldots,z_n')$ of the polydisk (4.20). In view of the ordinary Cauchy integral formula for holomorphic functions of one complex variable we have

$$w_k(z_1',\ldots,z_n') = \frac{1}{2\pi i} \int_{|z_1-z_{o1}|=r_1} \frac{w_k(z_1,z_2',\ldots,z_n')}{z_1 - z_1'} dz_1.$$

On the other hand we have

$$w_k(z_1,z_2',\ldots,z_n') = \frac{1}{2\pi i} \int_{|z_2-z_{o2}|=r_2} \frac{w_k(z_1,z_2,z_3',\ldots,z_n')}{z_2 - z_2'} dz_2$$

and so on. Substituting these formulas into each other, one obtains, finally, the n-dimensional Cauchy integral formula

$$w_k(z_1,\ldots,z_n) \tag{4.21}$$

$$= \frac{1}{(2\pi i)^n} \int_{|z_1-z_{o1}|=r_1} \cdots \int_{|z_n-z_{on}|=r_n} \frac{w_k(z_1,\ldots,z_n)}{(z_1-z_1')\ldots(z_n-z_n')} dz_1\ldots dz_n .$$

From this representation formula it is easy to derive the completeness of H_s in the case of several complex variables: Regard a Cauchy sequence of elements belonging to H_s and represent each element by the n-dimensional Cauchy integral formula (4.21). On the other hand, convergence in H_s means uniform convergence. Therefore, carrying out the limiting process under the integral sign, the limit of the given Cauchy sequence is representable by the Cauchy integral formula, too. The limit itself turns out, consequently, to be holomorphic (that way Weierstrass' convergence theorem has been proved for holomorphic functions of several complex variables).

Since each H_s is injected into each $H_{s'}$, if $s' < s$, the family H_s of the Banach spaces forms a scale in the sense of section 2.2. In order to apply the theorem about the solution of initial value problems in scales of Banach spaces (cf. section 3.5.), we must interpret the right-hand side of the system (4.17) of differential equations as mapping of the scale H_s, $0 < s < s_o$, into itself and verify that the conditions (I), (II), (III) (cf. 3.2.) are satisfied. First take into account that we look for solutions $w = w(t,z)$ of (4.17) that depend holomorphically on z for each fixed t, $0 \leq t \leq T$. Therefore the vector $w(t,z)$ may be interpreted as element of

$$\{t : 0 \leq t \leq T\} \times H_s$$

provided it is defined and continuous in $\overline{G_{z,s}}$ (and depends holomorphically on z in $G_{z,s}$). The vector $w(t,z)$ belongs even to the set

$$\{t : 0 \leq t \leq T\} \times \{w \in H_s : \|w\|_s \leq R\} \tag{4.22}$$

introduced in the case of an arbitrary scale of Banach spaces B_s by (3.15) in section 3.2. provided everywhere the inequality

$$|w(t,z)| \leq R$$

is satisfied.

On the other hand we assumed at the end of section 4.2. that the polydisk (4.18) is contained in G_w. Thus the composite functions defined by

$$a_{ik}^{(j)}(t,z,w(t,z)), \quad a_o^{(j)}(t,z,w(t,z)) \tag{4.23}$$

exist for each $w = w(t,z)$ belonging to the set (4.22). Since the superposition of holomorphic functions is holomorphic again, the composite functions (4.23) depend holomorphically on z. Since the w_k themselves depend holomorphically on each z_i by assumption, the derivatives $\frac{\partial w_k}{\partial z_i}$ are holomorphic in $G_{z,s}$, but they are not necessarily continuous in

$\overline{G_{z,s}}$. Thus for any $w = w(t,z)$ belonging to (4.22) the same is true for the right-hand side of (4.17). In any case the right-hand sides of (4.17) are continuous in $\overline{G_{z,s'}}$ for each fixed t if only $0 < s' < s$. In this way we have proved that the right-hand side of (4.17) may be interpreted as mapping of the set (4.22) into $H_{s'}$.

However, condition (I) is not yet confirmed by this. In order to show that condition (I) is satisfied completely we must prove, in addition, that the mapping of the set (4.22) into $H_{s'}$ defined by the right-hand side of the system (4.17) is continuous. For this end we deduce some estimates which are based on the n-dimensional Cauchy integral formula (4.21). Differentiating with respect to z_i', from formula (4.21) one obtains

$$\frac{\partial w_k}{\partial z_i'}(z_1',\dots,z_n')$$

$$= \frac{1}{(2\pi i)^n} \int_{|z_1-z_{o1}|=r_1} \cdots$$

$$\cdots \int_{|z_n-z_{on}|=r_n} \frac{w(z_1,\dots,z_n)}{(z_1-z_1')\dots(z_i-z_i')^2\dots(z_n-z_n')} dz_1\dots dz_n.$$

Substituting $z_j' = z_{oj}$ and estimating the integral, it follows at the point $(z_{o1}, \dots, z_{on}) = z_o$ the estimate

$$\left|\frac{\partial w_k}{\partial z_i}(z_o)\right| \leq \frac{1}{r_i}\|w\|_s \tag{4.24}$$

because the length of the path of integration in each z_j-plane is equal to $2\pi r_j$.

Denote the constant entering into (2.5) by c. Thus we have

$$\text{dist}\,(G_{z,s'}, \partial G_{z,s}) \geq c(s - s') \tag{4.25}$$

if $0 < s' < s$. Assume that the point z_o belongs even to $G_{z,s'}$. Regard the polydisk (4.20) with $r_1 = \dots = r_n = r$. Then the distance of an arbitrary point z of this polydisk from its centre z_o is not larger than than

$$d(z,z_o) = \left(\sum_{j=1}^{n} |z_j - z_{oj}|^2\right)^{\frac{1}{2}} \leq \sqrt{n}\, r.$$

Therefore in view of (4.26) the whole polydisk (4.20) with $r_1 = \dots$ $\dots = r_n = r$ is contained in $G_{z,s}$ if

$$r = \frac{c(s - s')}{\sqrt{n}}.$$

Substituting this expression into (4.24), where z_o is an arbitrary point of $G_{z,s'}$ and k varies from 1 to m, we get[1])

$$\left\|\frac{\partial w}{\partial z_i}\right\|_{s'} \leq \frac{\sqrt{n}}{c}\frac{1}{s' - s}\|w\|_s. \tag{4.26}$$

This inequality allows us to derive some necessary estimates on the mapping defined by the right-hand side of (4.17) in the set (4.22). For this end we regard two arbitrary elements (t', w') and (t", w") belonging to the set (4.22), where (t', w') is chosen fixedly and (t", w") varies in a certain neighbourhood of (t', w'). Substituting these two elements into the right-hand side of (4.17) and subtracting the arising expressions from each other, one obtains

$$\sum_{i=1}^{n}\sum_{j=1}^{m}\left(a_{ik}^{(j)}(t",z,w")\frac{\partial w"_k}{\partial z_i} - a_{ik}^{(j)}(t',z,w')\frac{\partial w'_k}{\partial z_i}\right)$$

$$+\left[a_o^{(j)}(t",z,w") - a_o^{(j)}(t',z,w')\right]. \tag{4.27}$$

We rewrite the first term of this difference in the form

$$\sum_{i=1}^{n}\sum_{j=1}^{m}\left[a_{ik}^{(j)}(t",z,w") - a_{ik}^{(j)}(t',z,w')\right]\frac{\partial w"_k}{\partial z_i}$$

$$+\sum_{i=1}^{n}\sum_{j=1}^{m} a_{ik}^{(j)}(t',z,w')\left(\frac{\partial w"_k}{\partial z_i} - \frac{\partial w'_k}{\partial z_i}\right). \tag{4.28}$$

Further, the coefficients of the derivatives $\frac{\partial w"_k}{\partial z_i}$ in the last expression are rewritten as

$$\left[a_{ik}^{(j)}(t",z,w") - a_{ik}^{(j)}(t",z,w')\right]$$

$$+\left[a_{ik}^{(j)}(t",z,w') - a_{ik}^{(j)}(t',z,s')\right]. \tag{4.29}$$

Analogously we rewrite the second term of (4.27).

In order to estimate the difference (4.27), we start with an estimate of the second term in (4.29). Define

$$S = \sup_{z\in G_{z,s}}\left|a_{ik}^{(j)}(t",z,w') - a_{ik}^{(j)}(t',z,w')\right|.$$

and take into account that $w' = w'(z)$ is defined and continuous in the compact set $\overline{G_{z,s}}$. Thus the composite functions $a_{ik}^{(j)}(t,z,w'(z))$ are uniformly continuous if z runs in $\overline{G_{z,s}}$, whereas t runs in a closed interval. Therefore, S turns out to be arbitrarily small if $|t" - t'|$ is sufficiently small. The same statement is true if S is defined with $a_o^{(j)}$

---

[1]) Since $w = w(z_1,\dots,z_n)$ depends (holomorphically) on n complex variables $z_1, \dots, z_n$ we denote the ordinary complex derivative with respect to z_i with $\partial w/\partial z_i$. It coincides with the partial complex derivative with respect to z_i, whereas the partial complex derivative with respect to $\bar{z}_i$, i.e., $\partial w/\partial \bar{z}_i$, is equal to zero.

instead of $a_{ik}^{(j)}$. In view of the Lipschitz-continuity of the $a_{ik}^{(j)}$ the first term in (4.29) can be estimated by

$$\left|a_{ik}^{(j)}(t'',z,w'') - a_{ik}^{(j)}(t'',z,w')\right| \leq l \sum_{j=1}^{m} |w_j'' - w_j'|.$$

Consequently, the module of the whole expression (4.29) is not larger than

$$lm\|w'' - w'\|_s + S. \tag{4.30}$$

In order to apply this bound for estimating the expression (4.28), we need further bounds for the derivatives appearing in (4.28). Since the pair (t", w") belongs to (4.22) we have $\|w''\|_s \leq R$ and, therefore, in view of (4.26) we obtain

$$\left\|\frac{\partial w''}{\partial z_i}\right\|_{s'} \leq \frac{\sqrt{n}}{c} \frac{R}{s - s'}. \tag{4.31}$$

Applying (4.26) to $w = w' - w''$, one gets, analogously,

$$\left\|\frac{\partial w''}{\partial z_i} - \frac{\partial w'}{\partial z_i}\right\|_{s'} \leq \frac{\sqrt{n}}{c} \frac{1}{s - s'}\|w'' - w'\|_s. \tag{4.32}$$

Taking into account that A is a bound of the modules of the coefficients, the expressions (4.30), (4.31) and (4.32) yield the following bound for the s'-norm of (4.28):

$$nm(lm\|w'' - w'\|_s + S)\frac{\sqrt{n}}{c} \frac{R}{s - s'} + nm\, A \frac{\sqrt{n}}{c} \frac{1}{s' - s}\|w'' - w'\|_s.$$

In this way we have estimated the s'-norm of the first term of (4.27), whereas (4.30) is a bound for the s-norm and, consequently, also for the s'-norm of the second term in (4.27). Therefore the s'-norm of (4.27) does not exceed the bound

$$\frac{mn\sqrt{n}}{c} \frac{1}{s - s'}((lmR + A)\|w'' - w'\|_s + SR) + lm\|w'' - w'\|_s + S. \tag{4.33}$$

Since S is arbitrarily small if $|t'' - t'|$ is sufficiently small the expression (4.33) shows, first, that the right-hand side of (4.17) defines a continuous mapping of the set (4.22) into $H_{s'}$. Thus condition (I) is satisfied. On the other hand we have $S = 0$ if $t'' = t'$.

Taking into account that

$$1 = \frac{s}{s} < \frac{s_0}{s - s'}$$

if $0 < s' < s$, in the case $S = 0$ the quantity (4.33) may be estimated by

$$\frac{C}{s - s'}\|w'' - w'\|_s,$$

where

$$C = \frac{mn\sqrt{n}}{c}(lmR + A) + lms_0.$$

Thus the right-hand sides of (4.17) satisfy condition (III), too.

It remains to show that condition (II) is also satisfied. Substituting the initial values (4.19) into the right-hand sides of (4.17), we obtain $a_o^{(j)}(t,z,0)$. Since

$$A = \frac{As_o}{s_o} < \frac{As_o}{s_o - s}$$

if $0 < s < s_o$, condition (II) is satisfied with $K = As_o$.

Since all three conditions (I), (II), (III) are satisfied, the theorem on the solution of initial value problems in scales of Banach spaces formulated in section 3.5. may be applied. Therefore the initial value problem (4.17), (4.19) is solvable by successive approximations. The constructed solution $w = w(t,z)$ belongs to H_s if $0 \leq t < a_*(s_o-s)$. Hence $w = w(t,z)$ turns out to exist in $G_{z,s}$ and to depend holomorphically on z if only t belongs to the semi-open interval $0 \leq t < a_*(s_o-s)$. That way the classical Cauchy-Kovalevskaya theorem formulated in section 4.1. has been proved completely.

4.4. A uniqueness theorem

In the preceding section we have proved that the quasilinear first order system (4.17) possesses a solution with identically vanishing initial values. The constructed solution turns out to depend holomorphically on z for each t. Therefore the solution belongs to the scale H_s of holomorphic functions for every t. On the other hand, the uniqueness theorem formulated in section 3.7. shows that the solution of the initial value problem in question is uniquely determined in the scale $H_{s'}$, i.e., there exists only one solution depending holomorphically on z for each t. In view of the third step in 4.2. the same is true in the case of arbitrary first order systems. Since systems of the order k are equivalent to first order systems (cf. the second step in 4.2.) there exists only one solution to the initial value problem (4.4), (4.12) that depends holomorphically on z for each t. Finally, in view of the first step in 4.2. the initial value problem (4.4), (4.5) with arbitrary holomorphic initial functions $\phi^{(o)}, \ldots, \phi^{(k-1)}$ possesses a uniquely determined solution being a holomorphic function in z for each t. In this way the following uniqueness theorem has been proved:

Theorem. The initial value problem (4.4), (4.5) possesses only one solution depending holomorphically on $z_1, \ldots, z_n$ for each t.

4.5. A real variant of the classical Cauchy-Kovalevskaya theorem

This section consists in two parts. Starting from a given system with complex differentiations, in the first part we shall interpret the classical Cauchy-Kovalevskaya theorem as statement for systems with real differentiations. On the contrary, the second part starts from certain real systems and will solve them by using the classical Cauchy-Kovalevskaya theorem.

I. Once more regard the initial value problem (4.4), (4.5), where the right-hand side f and the initial vectors $\phi^{(0)}, \phi^{(1)}, \ldots, \phi^{(k-1)}$ satisfy the conditions formulated in section 4.1. Suppose now that the point $x_0 = (x_{01}, \ldots, x_{0n})$ with real components x_{0i} belongs to G_z. Suppose further that the closed polycylinder

$$\{z = (z_1, \ldots, z_n) : |z_i - x_{0i}| \leq r_i^!, \ i = 1, \ldots, n\} \tag{4.34}$$

is contained in G_z, where all $r_i^!$ are positive (this implies that a larger open polycylinder centred at x_0 is contained in G_z, too). Then in view of the classical Cauchy-Kovalevskaya theorem in some t-interval the initial value problem in question possesses a solution in the polycylinder (4.34) depending holomorphically on $z_1, \ldots, z_n$. On the other hand, we know that in polycylinders holomorphic functions may be represented by power series, i.e.,

$$w(z_1, \ldots, z_n) = \sum_{i_1,\ldots,i_n} a_{i_1\ldots i_n}(t)(z_1 - x_{01})^{i_1} \ldots (z_n - x_{0n})^{i_n} \tag{4.35}$$

(the series is convergent even in a larger open polycylindric domain). It is well-known, further, that series of this kind may be differentiated term by term.

Now we regard the solution (4.35) only for real values of the variables $z_1, \ldots, z_n$, i.e., we consider the restriction of $w = w(z_1,\ldots,z_n)$ to the interval

$$\{x = (x_1, \ldots, x_n) : |x_i - x_{0i}| \leq r_i^!, \ i = 1, \ldots, n\} \tag{4.36}$$

belonging to the real n-dimensional Euclidian space $\mathbb{R}^n$. In this subset (4.36) of (4.34) the solution is given, of cause, by the power series

$$w(x_1,\ldots,x_n) = \sum_{i_1,\ldots,i_n} a_{i_1\ldots i_n}(t)(x_1 - x_{01})^{i_1} \ldots (x_n - x_{0n})^{i_n}, \tag{4.37}$$

which is a power series in real variables. The derivatives of such series with respect to the real variables also may be calculated by term-by-term differentiation. This is true also in the case that the coefficients $a_{i_1\ldots i_n}(t)$ are vectors with complex-valued components. Comparing the derivatives of (4.35) with those of (4.37), we see that the real derivatives of the function (4.37) coincide with the restrictions of the corresponding complex derivatives of the function (4.35) to the subset (4.36). Restricting the variable $z = (z_1, \ldots, z_n)$ to the variable $x = (x_1, \ldots, x_n)$, one obtains therefore that the components $w_j = w_j(x_1,\ldots,x_n)$ of $w = (w_1, \ldots, w_m)$ satisfy the differential equations

$$\frac{\partial^k w_j}{\partial t^k} = f_j(t,x_1,\ldots,x_n,w_1,\ldots,w_m,\frac{\partial w_1}{\partial t},\ldots,\frac{\partial^k w_m}{\partial x_n^k}), \tag{4.38}$$

$j = 1, \ldots, m$, instead of the complex system (4.2). Although the right-hand sides f_j may be complex-valued, the last system (4.38) may be interpreted as real one because all differentiations are carried out with respect to real variables. The initial values of the solution $w = w(x)$ of the system (4.38) are obtained by restricting the complex vector z in the initial values $\phi^{(0)}(z), \ldots, \phi^{(k-1)}(z)$ to the real vector x.

II. Conversely suppose that there is given such a real system (4.38) and initial vectors $\phi^{(0)}(x), \ldots, \phi^{(k-1)}(x)$ (where the right-hand sides and the initial vectors may be complex-valued). Suppose, however, that the right-hand sides possess power series representations with respect to $x_1, \ldots, x_n, w_1, \ldots, w_m$ and further with respect to all derivatives entering into the right-hand sides. Suppose that the closed interval (4.36) belongs to that domain in which the right-hand sides are representable as power series in $(x_1 - x_{01}), \ldots, (x_n - x_{0n})$.

On the other hand, we know from the complex function theory that a power series in x_i which is convergent for each x_i with $|x_i - x_{oi}| < \rho$ converges for each complex z_i with $|z_i - x_{oi}| < \rho$. Therefore such a power series can be extended automatically to a holomorphic function defined in a complex domain. We may assume, consequently, that the given right-hand sides are defined for complex z_i instead of the real x_i. Analogously, the right-hand sides are defined for complex values of the w_j and of their derivatives. Finally in the same way the given initial vectors $\phi^{(0)}(x), \ldots, \phi^{(k-1)}(x)$ are extended into the polycylinder (4.34). That way we have proved that the classical Cauchy-Kovalevskaya theorem is applicable. Therefore there exists a solution $w = w(t,z)$ of the complex extension of the given (real) initial value problem. Again restricting the complex variable z to the real variable x, we get the solution of the given real problem, i.e., the following theorem has been proved:

Theorem 1. The initial value problem

$$w(0,x) = \phi^{(0)}(x),$$

$$\frac{\partial w}{\partial t}(0,x) = \phi^{(1)}(x),$$

$$\vdots$$

$$\vdots \qquad (4.39)$$

$$\frac{\partial^{k-1} w}{\partial t^{k-1}}(0,x) = \phi^{(k-1)}(x)$$

for the real system (4.38) possesses a solution $w = w(t,x)$ representable as power series in $x_1 - x_{o1}, \ldots, x_n - x_{on}$ for each t belonging to some t-interval.

In order to prove the uniqueness of the investigated initial value problem in the case of the real system (4.38), we assume that there exists a second solution $\tilde{w} = \tilde{w}(t,x)$ of the same problem. Suppose further that $\tilde{w}(t,x)$ possesses power series representations in x for each t. Replacing x by z, we obtain a continuation $\tilde{w} = \tilde{w}(t,z)$ of this function depending holomorphically on z for each t. In view of I. the derivatives of $w = w(t,x)$ are the restrictions of the corresponding complex derivatives to $z = x$. Since $\tilde{w} = \tilde{w}(t,x)$ satisfies the system (4.38) (by assumption), the extended function $\tilde{w} = \tilde{w}(t,z)$ satisfies the complex equations (4.2) for $z = x$. On the other hand we know that a holomorphic function must vanish identically if it vanishes on the real axes. Therefore the complex differential equations (4.2) must be satisfied by $\tilde{w} = \tilde{w}(t,z)$ everywhere. Thus $w = w(t,z)$ and $\tilde{w} = \tilde{w}(t,z)$ are two solutions of (4.2) possessing the same initial vectors $\phi^{(o)}(z),\ldots,\phi^{(k-1)}(z)$. Applying the uniqueness theorem 4.4., we conclude that the two solutions must be identical. Therefore the following uniqueness theorem has been proved:

<u>Theorem 2.</u> The initial value problem (4.38), (4.39) possesses only one solution $w = w(t,x)$ being representable into a power series with respect to x for each t.

This theorem does not exclude the possibility of the existence of further solutions to (4.38), (4.39) being not necessarily representable by power-series in x but possessing derivatives up to the order k at least. The next chapter deals with the so-called Holmgren theorem in view of which every solution to (4.38), (4.39) must be representable by power-series with respect to $x_1, \ldots, x_n$ provided the right-hand sides f_j as well as the initial vectors $\phi^{(o)}, \ldots, \phi^{(k-1)}$ possess the same property.

5. THE HOLMGREN THEOREM

5.1. Statement of the problem

Theorem 1 in section 4.5. shows that the initial value problem (4.39) to the system (4.38) of real differential equations is solvable by power series in $x_1, \ldots, x_n$ (cf. (4.37)) provided the initial functions $\phi^{(o)}, \ldots, \phi^{(k-1)}$ are power series in $x_1, \ldots, x_n$ and the right-hand sides $f_1, \ldots, f_m$ are power series in $x_1, \ldots, x_n, w_1, \ldots, w_m$ and all derivatives entering into the right-hand sides. The second theorem of the same section states that there exists only one solution being representable as power-series in $x_1, \ldots, x_n$. Both theorems are applicable, particularly, to linear first order systems (the right-hand sides of such systems are linear combinations of the sought functions and their derivatives with respect to the x_i). We interpret such systems as real ones since the differentiations are carried out with respect to real variables indifferently whether the coefficients as well as the desired functions are real- or complex-valued. Splitting up complex-valued functions w_j, complex-valued initial functions $\phi^{(o)}, \ldots, \phi^{(k-1)}$ and complex-valued coefficients $a_{ik}^{(j)}$, $a_k^{(j)}$, $a_o^{(j)}$ into their real and imaginary parts, without any loss of generality we may assume, however, that the above-mentioned functions are real-valued. Therefore we may restrict ourselves to systems

$$\frac{\partial u_j}{\partial t} = \sum_{i=1}^{n} \sum_{k=1}^{m} a_{ik}^{(j)}(t,x)\frac{\partial u_k}{\partial x_i} + \sum_{k=1}^{m} a_k^{(j)}(t,x)u_k + a_o^{(j)}(t,x), \tag{5.1}$$

$j = 1, \ldots, m$, with real-valued coefficients and real-valued functions u_k looked for. Regard the initial value problem

$$u(0,x) = \phi(x), \tag{5.2}$$

where $u = (u_1, \ldots, u_m)$ and the components of the inital vector ϕ are real-valued, too. The Holmgren theorem is the following statement:

Theorem. Suppose that the coefficients of the system (5.1) and the initial vector are representable as power series with respect to all their variables. Then each continuously differentiable solution to the initial value problem is representable as power series in $x_1, \ldots$ x_n. Thus the solution to the initial value problem is uniquely determined.

In view of the theorems 1 and 2 of 4.5. the initial value problem in question possesses at least one solution. Thus it remains to prove that the difference of two solutions vanishes identically. On the other hand, the difference of two solutions to the system (5.1) (which is inhomogeneous, in general) satisfies the homogeneous system

$$\frac{\partial u_j}{\partial t} = \sum_{i=1}^{n} \sum_{k=1}^{m} a_{ik}^{(j)}(t,x)\frac{\partial u_k}{\partial x_i} + \sum_{k=1}^{m} a_k^{(j)}(t,x)u_k, \tag{5.3}$$

$j = 1, \ldots, m$. Further the initial values of the difference of two solutions vanish identically. Hence, the Holmgren theorem will be proved if we show that each continuously differentiable solution $u = u(t,x)$ to the homogeneous system (5.3) satisfying the homogeneous initial condition

$$u(0,x) = (0, \ldots, 0) \tag{5.4}$$

must vanish identically at each point x for each t.

5.2. Proof of the Holmgren theorem

For short we carry out the proof of the Holmgren theorem only in the case of the differential equation

$$\frac{\partial u}{\partial t} = a(t,x)\frac{\partial u}{\partial x} + b(t,x)u, \tag{5.5}$$

where the desired solution $u = u(t,x)$ as well as the coefficients $a(t,x)$, $b(t,x)$ are complex-valued, whereas the independent variable x is real. Such systems (5.5) are special systems of type (5.3) with $m = 2$ and $n = 1$ (cf. the remark 5.3.1.). In our case the initial condition (5.4) can be written as

$$u(0,x) = 0 \tag{5.6}$$

for each x. Corresponding to the assumptions on the coefficients of the system (5.1) we suppose that $a(t,x)$ and $b(t,x)$ are power series in t and x.

For the sake of clarity subdivide the proof of the Holmgren theorem into six parts: In the first part a given solution to (5.5) which is defined for non-negative values of t will be extended to negative t. After that a transformation of the variable t will be carried out. The third part will interpret a given solution to (5.5) as functional for which a differential equation will be deduced in the fourth part. Then the theory of dual scales (cf. 2.4.) will be applied. In the last part finally, it will be shown that the regarded solution must vanish identically (the following proof is derived from F. Treves' book [61]).

I. Extension of a given solution

Let $u = u(t,x)$ be a given continuously differentiable solution to (5.5) satisfying the initial condition (5.6). Assume that $u(t,x)$ is defined for each t of the interval $0 \leq t \leq T$ if x belongs to an open interval of the x-axis. Now regard any point x_0 of this interval. Without any loss of generality we may assume that $x_0 = 0$. Thus $u = u(t,x)$ is defined in the rectangle

$$\{t : 0 \leqq t \leqq T\} \times \{x : |x| < r\},$$

where r is suitably chosen. In view of (5.6) both initial functions $u(0,x)$ and $\frac{\partial u}{\partial x}(0,x)$ vanish identically. Thus the differential equation (5.5) shows that $\frac{\partial u}{\partial t}$ vanishes indentically if t equals to zero.

Until now the function $u = u(t,x)$ is defined only for non-negative values of t. Defining it by

$$u(t,x) = 0$$

for $t < 0$, the extended function $u = u(t,x)$ turns out to be a continuously differentiable solution to (5.5) at least in the enlarged rectangle

$$\{t : - T \leqq t \leqq + T\} \times \{x : |x| < r\}, \tag{5.7}$$

where T and r are assumed to be sufficiently small so that $a(t,x)$ and $b(t,x)$ are given in the whole interval (5.7).

II. Transformation of the variables

Now introduce new variables (t',x) instead of (t,x) by the formula

$$t' = t + x^2, \quad \text{i.e.} \quad t = t' - x^2. \tag{5.8}$$

This transformation defines a 1-1-mapping of a neighbourhood of the point (0,0) of the t,x-plane onto a neighbourhood of the point (0,0) in the t',x-plane since the Jacobian

$$\frac{\partial(t',x)}{\partial(t,x)} = \begin{vmatrix} 1 & 2x \\ 0 & 1 \end{vmatrix} = 1$$

is different from zero everywhere. Denote the function $u = u(t,x)$ interpreted as function of the new coordinates (t',x) by $\tilde{u} = \tilde{u}(t',x)$, i.e.,

$$u(t,x) = \tilde{u}(t',x) = \tilde{u}(t + x^2, x).$$

Therefore one obtains

$$\frac{\partial u}{\partial t}(t,x) = \frac{\partial \tilde{u}}{\partial t'}(t',x),$$

$$\frac{\partial u}{\partial x}(t,x) = \frac{\partial \tilde{u}}{\partial t'}(t',x)\cdot 2x + \frac{\partial \tilde{u}}{\partial x}(t',x),$$

and the differential equation (5.5) is transformed into

$$\frac{\partial \tilde{u}}{\partial t'} = \tilde{a}(t',x)\frac{\partial \tilde{u}}{\partial x} + \tilde{b}(t',x)\tilde{u}, \tag{5.9}$$

where

$$\tilde{a}(t',x) = \frac{a(t'-x^2,x)}{1 - 2xa(t'-x^2,x)},$$
$$\tag{5.10}$$
$$\tilde{b}(t',x) = \frac{b(t'-x^2,x)}{1 - 2xa(t'-x^2,x)}.$$

Since a(t,x) is locally bounded the denominator in (5.9) is different from zero in a sufficiently small neighbourhood of (0,0). The formulas (5.10) show immediately that the coefficients $\tilde{a}(t',x)$ and $\tilde{b}(t',x)$ are representable by power-series, too. The transformation (5.8) maps the half-plane $\{(t,x) : t < 0\}$ onto the set of all points lying below the parabola $t' = x^2$, i.e.,

$$\{(t',x) : t' < x^2\}. \tag{5.11}$$

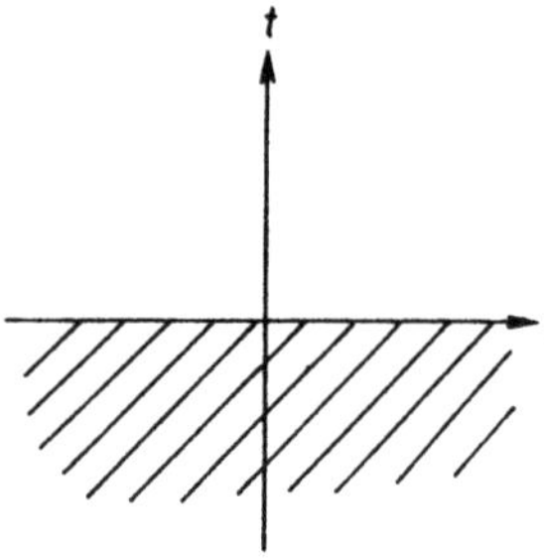

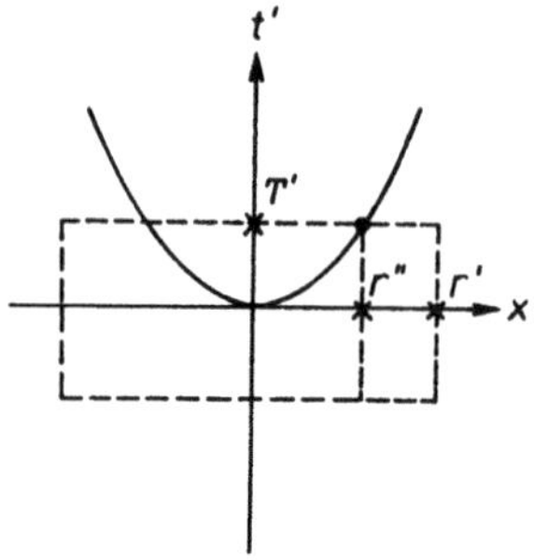

Since u = u(t,x) vanishes identically if t < 0, the function $\tilde{u} = \tilde{u}(t',x)$ vanishes also identically in all points of the regarded neighbourhood of (0,0) that belong to the set (5.11).

Choose the neighbourhood of (0,0) in the t',x-plane as rectangle

$$\{t' : |t'| < T'\} \times \{x : |x| < r'\}, \tag{5.12}$$

where T' and r' are sufficiently small. If T' is small enough, then there exists a number r'', $0 < r'' < r'$, such that $\tilde{u}(t',x)$ can be different from zero only if $|x| \leq r''$ (in other words, the function $\tilde{u} = \tilde{u}(t',x)$ interpreted as function of x possesses a compact support in the interval $\{x : |x| < r'\}$ for each t'). That is the reason why it is advantageous to introduce the new variable t'.

III. Interpretation of the solution as a functional

In accordance with the preceding section II. we may assume that $\tilde{u} = \tilde{u}(t',x)$ is a given solution defined in the rectangle (5.12), where $\tilde{u} = \tilde{u}(t',x)$ vanishes identically with respect to t' if only $r'' \leq |x| \leq r'$. Now regard a family of subdomains G'_s of the interval $G' = \{x : |x| < r'\}$, namely

$$G'_s = \{x : |x| < r'' + s(r' - r'')\},$$

where $0 < s < 1$.

Then the G'_s are a possible family of subdomains of the domain G' in the sense of section 2.2., i.e., the corresponding three conditions formulated in section 2.2. are satisfied. Especially we have

$$\text{dist}\,(G_{s'}, \partial G_s) \geq (r' - r'')(s - s')$$

if $0 < s' < s < 1$ so that the constant entering into the condition b) equals to $r' - r''$. Now define a scale of Banach spaces H'_s, $0 < s < 1$, consisting of (complex-valued) functions $h = h(x)$ with the following properties:

a) They are representable as power series in the real variable x if x belongs to G_s (the coefficients of the power series are complex, in general).

b) The complex extensions $h(z)$ of the power series $h(x)$ are continuous even in the closed disk

$$\{z : |z| \leq r'' + s(r' - r'')\}$$

(note that the power-series $h(x)$ converges for every complex z with $|z| < r'' + s(r' - r'')$, cf. section 4.5., part I.). Introduce a norm into H'_s by setting

$$\|h\|_s = \sup_{|z|<r''+s(r'-r'')} |h(z)|,$$

i.e., the norm is defined by the supremum of the module of the complex extension $h = h(z)$ of $h = h(x)$.

Take any Cauchy sequence in H'_s that is formed by functions $h_n = h_n(x)$, $n = 1, 2, \ldots$ In view of the definiton of the norm in H'_s the holomorphic extensions $h_n = h_n(z)$ form a Cauchy sequence, too. Since the limit of a uniformly convergent sequence of holomorphic functions is again holomorphic, the limit function $h = h(z)$ of the holomorphic extensions $h_n = h_n(z)$ is a power series in z. Therefore, the restriction $h = h(x)$ of the limit function to the real axis is a power series in x. Thus the spaces H'_s turn out to be (closed) subspaces of the corresponding spaces H_s of the holomorphic extensions (cf. 2.2.).

The ordinary complex differentiation $\frac{d}{dz}$, furthermore, is a bounded operator mapping H_s into $H_{s'}$, $0 < s' < s < 1$, where its norm can be estimated by

$$\frac{\text{const}}{s - s'}$$

(cf. 2.3.). Since the restriction of $\frac{dh}{dz}(z)$ to real values $z = x$ is equal to the real derivative $\frac{dh}{dx}(x)$ (cf. 4.5., I.), the following statement is true:

The real differentiation $\frac{d}{dx}$ is a bounded operator mapping H'_s into $H'_{s'}$, $0 < s' < s$, where its norm does not exceed the number

$$\frac{\text{const}}{s - s'} \tag{5.13}$$

with a constant not depending on s, s'.

Take any element $h = h(x)$ belonging to $H_s^!$ and define

$$U(t')[h] = \int_{G_s^!} \tilde{u}(t',x)\, h(x)\, dx, \tag{5.14}$$

where $\tilde{u} = \tilde{u}(t',x)$ is the given solution to (5.9). Since $\tilde{u} = \tilde{u}(t',x)$ is defined in the rectangle (5.12) the right-hand side of (5.14) defines a (complex-valued) functional $U(t')$ on $H_s^!$ for each t' with $0 \leqq t' < T'$, i.e., $U(t')$ belongs to the dual space $H_s^{!*}$ of $H_s^!$ if $0 \leqq t' < T'$. Especially we have $U(0) = 0$ because $\tilde{u}(0,x)$ vanishes identically.

IV. The functional U(t') as a solution of an ordinary differential equation

Since $\tilde{u} = \tilde{u}(t',x)$ is a solution to the partial differential equation (5.9) vanishing identically for $t' = 0$, we obtain

$$\tilde{u}(t',x) = \int_0^{t'} \left(\tilde{a}(\tau,x)\frac{\partial\tilde{u}}{\partial x}(\tau,x) + \tilde{b}(\tau,x)\,\tilde{u}(\tau,x)\right)d\tau .$$

Substituting this representation into (5.14) and altering the order of the integrations, one gets

$$U(t')[h] = \int_0^{t'} \left[\int_{G_s^!} \left(\tilde{a}(\tau,x)\,\frac{\partial\tilde{u}}{\partial x}(\tau,x) + \tilde{b}(\tau,x)\,\tilde{u}(\tau,x)\right)h(x)\,dx\right]d\tau .$$

In accordance with theorem 2 of section 1.4., the last formula yields

$$\frac{dU}{dt'}(t')[h] = \int_{G_s^!} \left(\tilde{a}(t',x)\,\frac{\partial\tilde{u}}{\partial x}(t',x) + \tilde{b}(t',x)\,\tilde{u}(t',x)\right)h(x)\,dx. \tag{5.15}$$

Now take into account that $\tilde{u} = \tilde{u}(t',x)$ possesses a compact support. Thus $\tilde{u} = \tilde{u}(t',\cdot)$ vanishes at the boundary of the interval $G_s^!$. Integrating the first term on the right-hand side of (5.15) by parts one obtains, therefore,

$$\begin{aligned} &\frac{dU}{dt'}(t')[h] \\ &= \int_{G_s^!} \tilde{u}(t',x)\left[\left(-\frac{\partial\tilde{a}}{\partial x}(t',x) + \tilde{b}(t',x)\right)h(x) - \tilde{a}(t',x)\,\frac{dh}{dx}(x)\right]dx. \end{aligned} \tag{5.16}$$

This representation formula holds also in the case that the derivative $\frac{dh}{dx}$ is unbounded at the boundary $\partial G_s^!$ of $G_s^!$ because the support of $u = u(t',x)$ is a compact subset of $G_s^!$. Denote the second factor of the integrand in the last equation by $A(t)h$, i.e., we define

$$\left(-\frac{\partial\tilde{a}}{\partial x}(t',\cdot) + \tilde{b}(t',\cdot)\right)h - \tilde{a}(t',\cdot)\frac{dh}{dx} = A(t')h.$$

Thus $A(t')$ is a linear operator defined on $H_s^!$. Since the coefficients $\tilde{a}$, $\tilde{b}$ depend on t', in general, this operator depends on t', too (this fact is expressed by the notation $A(t')$ for this operator). Without any loss of generality we may assume that the modules of

$$\tilde{a}(t',x),\ \tilde{b}(t',x) \text{ and } \frac{\partial \tilde{a}}{\partial x}(t',x)$$

are bounded by a common constant in the set (5.12). On the other hand, $\frac{d}{dx}$ is a bounded operator mapping H'_s into $H'_{s'}$ if $0 < s' < s < 1$, where its norm can be estimated by (5.13). Summarizing these considerations, we see that $A(t')$ is also a bounded operator mapping H'_s into $H'_{s'}$, where

$$\|A(t')\| \leq \frac{\text{const}}{s - s'}. \tag{5.17}$$

Note that the constant in the last inequality differs from that in (5.13), in general.

In view of the definition of the operator $A(t')$ the formula (5.16) may be rewritten as

$$\frac{dU}{dt'}(t')[h] = \int_{G'_s} \tilde{u}(t',x)\ A(t')\ h(x)\ dx.$$

Taking into account the definition (5.14) of the functional $U(t')$, the last equation yields

$$\frac{dU}{dt'}(t')[h] = U(t')[A(t')h]. \tag{5.18}$$

V. Application of the theory of dual scales

First we would like to remember that the support of $\tilde{u} = \tilde{u}(t',x)$ is contained in each $G'_{s'}$, $0 < s' < s$, so that the functional $U(t')$ on the right-hand side also may be interpreted as functional on $H'_{s'}$. This interpretation is essential because $A(t')h$ belongs to $H'_{s'}$ if h is an element belonging to H'_s. Since $A(t')$ maps H'_s into $H'_{s'}$ the adjoint operator $A^*(t')$ maps $H'^*_{s'}$ into H'^*_s if only $s' < s$. Using the definition (2.6) of adjoint operators, the right-hand side of (5.18) may be rewritten as

$$\left(A^*(t')U(t')\right)[h], \tag{5.19}$$

where the desired functional $U(t')$ belongs to H'^*_s. Since $H'_{s'} \supset H'_s$ we have $H'^*_{s'} \subset H'^*_s$ and, consequently, each functional belonging to $H'^*_{s'}$ belongs to H'^*_s, too. Thus the functional $\frac{dU}{dt}(t')$ on the left-hand side of (5.18) may be interpreted not only as element of $H'^*_{s'}$ but also of H'^*_s.

Replacing the right-hand side of (5.18) by (5.19), we see that the functional $U(t')$ defined by (5.14) is a solution to the differential equation

$$\frac{dU}{dt}(t') = A^*(t')U(t') \tag{5.20}$$

in the dual scale $H'^*_{s'}$. Applying the estimates (2.7) and (5.17), we get that the norm of the operator $A^*(t')$ interpreted as mapping from $H'^*_{s'}$ into H'^*_s may be estimated by

$$\|A^*(t')\| \leq \frac{\text{const}}{s - s'}. \tag{5.21}$$

Now return to the example at the end of section 2.4. Let $h = h(x)$ be any element of $H'_{s'}$. The mentioned example shows that the complex extension $h(z)$ of $h(x)$ may be approximated uniformly by polynomials $p(z)$ in the whole closed disk. Replacing the variable z by x, one obtains polynomials $p(x)$ in x (the coefficients of the Taylor series of $h(z)$ are complex, in general, so that $p(x)$ may be complex-valued on the real axis). On the other hand polynomials in x belong to H'_s. Since $\|h(z) - p(z)\|_{s'}$ is arbitrarily small the space H'_s, $s > s'$, turns out to be dense in $H'_{s'}$. By virtue of the theorem of the section 2.4. the dual spaces H'^*_s form a scale of Banach spaces, too.

Applying the theorem of 3.8., in view of (5.21) each initial value problem to the differential equation (5.20) is uniquely solvable.

A special solution to (5.20) is given by the functional $U(t')$ defined by (5.14).

As it was demonstrated at the end of the part III. of the present section, this functional $U(t')$ vanishes identically at the point $t' = 0$, i.e.,

$$U(0) = 0. \tag{5.22}$$

On the other hand, a special solution to the initial value problem (5.20), (5.22) is given by the zero functional at each point t', i.e., $U(t') = 0$ for every t'. Since the solution to (5.20), (5.22) is unique we obtain that the functional (5.14) must be equal to the zero functional $U(t') = 0$ on $H'^*_{s'}$ for each t'. Hence the relation

$$\int_{G'_{s'}} \tilde{u}(t',x)\, h(x)\, dx = 0 \tag{5.23}$$

must be satisfied for each t' if $h = h(x)$ is any element of $H'_{s'}$ (notice that the domain of integration in (5.15), i.e., G'_s, may be replaced by the smaller interval $G'_{s'}$ because the support of $u = u(t',x)$ is contained in $G'_{s'}$, too).

VI. The generating function of a zero functional of type (5.14)

Now we are going to prove that the generating function $\tilde{u} = \tilde{u}(t',x)$ must vanish identically in both variables t' and x provided the relation (5.23) holds for each element $h = h(x)$ of $H'_{s'}$. For this end we choose any positive number ε. Then regard the function $\tilde{u} = \tilde{u}(t',x)$ for fixed t' and arbitrary x in $\overline{G'_{s'}}$. In view of the Stone-Weierstrass theorem (cf., for instance, P. S. Aleksandrov [2]) there exists a polynomial $p(x)$ such that

$$|\tilde{u}(t',x) - p(x)| < \varepsilon \tag{5.24}$$

for each x of $G'_{s'}$ (if p_1 and p_2 are $\frac{\varepsilon}{2}$-approximations of Re $\tilde{u}$ and Im $\tilde{u}$ resp., then $p = p_1 + ip_2$ satisfies the inequality (5.24)). By virtue of (5.23) we have

$$\int_{G'_{s'}} \tilde{u}(t',x)\,\overline{p(x)}\,dx = 0$$

since $p = p(x)$ and $\overline{p} = \overline{p(x)}$ belong to $H'_{s'}$. The last relation shows that

$$\int_{G'_{s'}} |\tilde{u}(t',x)|^2 dx = \int_{G'_{s'}} \tilde{u}(t',x)\,\overline{(\tilde{u}(t',x))}\,dx$$

$$= \int_{G'_{s'}} u(t',x)\,\overline{(\tilde{u}(t',x) - p(x))}\,dx\ .$$

Taking into account the inequality (5.24), the last equality yields the estimate

$$\int_{G'_{s'}} |\tilde{u}(t',x)|^2 dx \leq \sup_{G'_{s'}} |\tilde{u}(t',x)| \cdot \varepsilon \cdot 2r'$$

since the length of the interval $G'_{s'}$ is less than 2r' in any case (concerning the meaning of r' see (5.12)). Choosing ε sufficiently small, the right-hand side of (5.25) turns out to be arbitrarily small. Therefore, in view of the continuity of $\tilde{u}$ the integrand in (5.25) must vanish identically. That way we have proved that $u = u(t',x)$ is equal to zero at each point of the rectangle (5.12).

Returning to the original variables t and x, we have proved the following statement:

Let $u = u(t,x)$ be any continuously differentiable solution to the initial value problem (5.5), (5.6) defined in an open x-interval for each t with $0 \leq t \leq T$. Then to each point x_0 of this x-interval there exists a neighbourhood in which $u = u(t,x)$ vanishes identically for sufficiently small t (the t-interval can be chosen uniformly for each x belonging to a compact subset of the x-interval mentioned above).

5.3. Further remarks on the classical Holmgren theorem

5.3.1. The proof of the Holmgren theorem has been carried out in the case of the special system (5.5), where the coefficients a(t,x), b(t,x) and the solution $u = u(t,x)$ looked for are complex-valued. In the case of m desired real-valued functions denote the vector $(u_1, \ldots, u_m)$ by u. For $n = 1$ the general system (5.3) may be rewritten as

$$\frac{\partial u}{\partial t} = a(t,x)\left(\frac{\partial u}{\partial x}\right)^{\text{transp}} + b(t,x)u^{\text{transp}}, \qquad (5.25)$$

where "transp" means the transpose, and a(t,x), b(t,x) are matrices depending on t and x. If only the spaces H'_s of 5.2., III. are replaced by analogous spaces of vector-valued functions $h(x) = (h_1(x), \ldots, h_m(x))$, all considerations of 5.2. remain true in the case of the vector equation (5.25). For an arbitrary n equation (5.25) is to be replaced by the still more general equation

$$\frac{\partial u}{\partial t} = \sum_{i=1}^{n} a_i(t,x)\left(\frac{\partial u}{\partial x_i}\right)^{\text{transp}} + b(t,x)u^{\text{transp}} \tag{5.26}$$

whose coefficients $a_i(t,x)$, $b(t,x)$ are matrices, again. Also in the case of the general linear system (5.26) the proof 5.2. is practicable because in the scale each of the n differentiations $\frac{\partial}{\partial x_i}$, $i = 1, \ldots, n$, is a bounded operator whose norm can be estimated by (5.13). The transformation (5.8) of the variable t is, however, to be replaced by

$$t' = t + \sum_{i=1}^{n} x_i^2. \tag{5.27}$$

Finally we would like to return to the scalar equation (5.5). If the coefficients $a(t,x)$, $b(t,x)$ as well as the desired solution $u = u(t,x)$ are only real-valued, the statement of the Holmgren theorem may be proved with the method of characteristics allowing us to reduce partial differential equations of first order for one real-valued function looked for to a system of ordinary differential equations. The corresponding uniqueness theorem holds even in the case of continuously differentiable coefficients (see, for instance, R. Courant and D. Hilbert [13]).

5.3.2. In view of the Cauchy-Kovalevskaya theorem (theorem 1 in 4.5.) the initial value problem (5.1), (5.2) is solvable provided the coefficients as well as the initial functions are power series. If the initial functions are not power series, the initial value problem is not necessarily solvable (even in the case that the coefficients are power series). The Holmgren theorem shows also in such cases that the solution is uniquely determined if there exists one at least.

5.3.3. The proof of the Holmgren theorem carried out in section 5.2. cannot be applied in the case of quasilinear differential equations. The reason is that in view of

$$\tilde{a}(t',x,\tilde{u})\frac{\partial \tilde{u}}{\partial x}(t',x)h(x)$$

$$= \frac{\partial}{\partial x}(\tilde{a}(t',x,\tilde{u})\ \tilde{u}(t',x)\ h(x)) - \frac{\partial \tilde{a}}{\partial x}(t',x,\tilde{u})\ \tilde{u}(t',x)\ h(x)$$

$$- \frac{\partial \tilde{a}}{\partial u}(t',x,u)\ \frac{\partial \tilde{u}}{\partial x}(t',x)\ \tilde{u}(t',x)\ h(x) - \tilde{a}(t',x,\tilde{u})\ \frac{dh}{dx}(x)$$

in the integrand of (5.16) the additional term

$$- \frac{\partial \tilde{a}}{\partial u}\ \frac{\partial \tilde{u}}{\partial x}\ \tilde{u}h$$

occurs.

5.3.4. Now regard the system (4.38) of order k in the special case that the right-hand sides depend linearly on

$$w_1, \ldots, w_m, \frac{\partial w_1}{\partial t}, \ldots, \frac{\partial^k w_m}{\partial x_n^k},$$

i.e., the right-hand sides of the equations (4.38) are linear combinations of the desired functions and their derivatives, where the coefficients of these linear combinations are power series in $t, x_1, \ldots, x_n$. Interpreting first order derivatives as new functions looked for (cf. (4.9), (4.10)), the order may be reduced step by step (see 4.2., second step). This reduction may be carried out with real derivatives, too. Finally one obtains a linear first order system for an enlarged set of desired functions. Applying the Holmgren theorem 5.1. to this system, one gets, consequently, the following Holmgren theorem for higher order systems:

Theorem. Suppose that the coefficients of a homogeneous linear system of type (4.38) are power series in $t, x_1, \ldots, x_n$. Suppose, further, that $w_1, \ldots, w_m$ is a given solution being k times continuously differentiable and satisfying homogeneous initial conditions, i.e.,

$$w(0,x) = 0$$

$$\frac{\partial w}{\partial t}(0,x) = 0$$

$$\vdots$$

$$\frac{\partial^{k-1} w}{\partial t^{k-1}}(0,x) = 0$$

(cf. (4.39)). Then the solution $w = w(t,x)$ must vanish idendically.

This theorem shows, for instance, that the solution $w = w(t,x)$ of any initial value problem (4.39) turns out to be a power series in $x_1, \ldots$ $\ldots, x_n$ for each t provided the coefficients of the system as well as the initial functions $\phi^{(0)}, \ldots, \phi^{(k-1)}$ are representable by power series.

5.3.5. Finally we would like to explain the basic idea of the original proof of Holmgren's theorem. To it we regard again the homogeneous differential equation (5.5) for a desired complex-valued function $u = u(t,x)$, where x is a real variable. Suppose again that $u = u(t,x)$ is defined if $t \geq 0$ and, furthermore, that the initial values $u(0,x)$ vanish identically. Without any loss of generality we assume that the

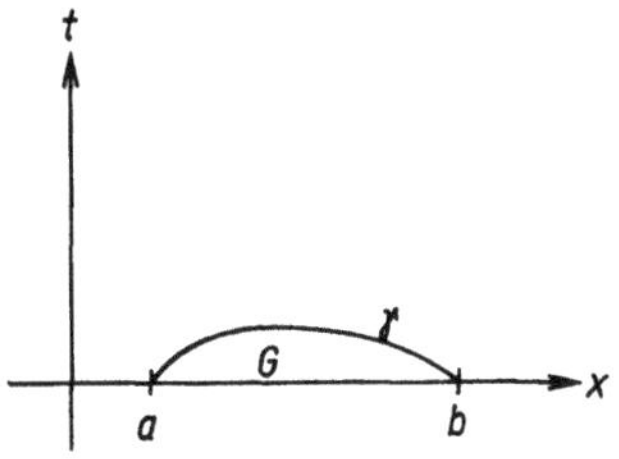

given solution is defined in an open interval of the x-axis containing the closed interval [a, b]. Now take a curve γ defined by $t = \psi(x)$, $a \leq x \leq b$, and contained in the upper half-plane $t > 0$ with the exception of its initial and ending points that coincide with the initial and ending points of the interval [a, b] (cf. the figure). Let G be the domain between the interval [a, b] and the curve γ. The boundary of G consists, consequently, of the interval [a, b] on the x-axis and the curve γ.

Now regard the expression

$$\frac{\partial}{\partial t}(uv) - \frac{\partial}{\partial x}(auv) = v\left[\frac{\partial u}{\partial t} - a\frac{\partial u}{\partial x} - bu\right] + u\left[\frac{\partial v}{\partial t} - a\frac{\partial v}{\partial x} - \frac{\partial a}{\partial x}v + bv\right]. \tag{5.28}$$

On the other hand, in view of Green's integral formula we have

$$\iint_G \left(\frac{\partial}{\partial t}(uv) - \frac{\partial}{\partial x}(auv)\right)dxdt = -\int_{\partial G} uvdx - \int_{\partial G} auvdt. \tag{5.29}$$

The first term on the right-hand side of (5.28) vanishes since u is a solution to the homogeneous differential equation (5.5). Now let v be a solution to the differential equation

$$\frac{\partial v}{\partial t} = a\frac{\partial v}{\partial x} + \frac{\partial a}{\partial x}v - bv + p(t,x), \tag{5.30}$$

where p(t,x) is an arbitrary polynomial in t and x. Then the second term on the right-hand side of (5.28) equals to $u \cdot p$ and, therefore, formula (5.29) passes into

$$\iint_G updxdt = -\int_{\partial G} uvdx - \int_{\partial G} auvdt. \tag{5.31}$$

Remember that u vanishes identically on the x-axis. Hence the path of integration of the integrals on the right-hand side of (5.31) may be reduced to the curve γ. These integrals over the curve γ vanishes, too, if v vanishes on γ. If γ is an analytic curve, then there exists a solution v to the differential equation (5.30) which vanishes identically on γ. Provided $\sup_{a \leq x \leq b} |\psi(x)|$ is small enough, the solution $v = v(t,x)$ exists in the whole domain G and, consequently, formula (5.31) may be applied. This formula yields

$$\iint_G updxdt = 0$$

for any polynomial p(t,x). This relation implies, however, that u vanishes identically (cf. 5.2., step VI).

By using Green's integral formula for higher order differential operators the last considerations may be applied immediately in the case of differential equations of higher order, too (cf. R. Courant and D. Hilbert [13], L. Bers, Fritz John, M. Schechter [9], D. L. Colton [12]).

5.4. A generalization of the Holmgren theorem

The classical Holmgren theorem states that each continuously differentiable solution to the homogeneous differential equation (5.3) satisfying the homogeneous initial condition (5.4) is everywhere equal to zero provided the coefficients $a_{ik}^{(j)}(t,x)$, $a_k^{(j)}(t,x)$ of the differential equation (5.3) are power series in t and x. Now we are going to prove that the same statement is true for more general coefficients of the differential equation[1]).

Choose, first, a real-valued function $\lambda = \lambda(x)$, $x = (x_1, \ldots, x_n)$, defined in a neighbourhood of the origin 0 of the x-space and satisfying the following conditions:

a) $\lambda(x)$ is a power-series in $x_1, \ldots, x_n$;
b) $\lambda(0) = 0$;
c) $\lambda = \lambda(x)$ possesses a (strict) minimum at the point 0, i.e., we have $\lambda(x) > 0$ for each $x \neq 0$.

Now regard a neighbourhood of the origin of the (n+1)-dimensional (t,x)-space. Introduce new variables (t',x) instead of (t,x) by setting

$$t' = t + \lambda(x). \tag{5.32}$$

Since the Jacobian is equal to

$$\frac{\partial(t',x)}{\partial(t,x)} = \begin{vmatrix} 1 & \frac{\partial\lambda}{\partial x_1} & \cdots & \frac{\partial\lambda}{\partial x_n} \\ 0 & 1 & \cdots & 0 \\ \cdot & \cdot & \cdot & \cdot \\ \cdot & \cdot & \cdot & \cdot \\ \cdot & \cdot & \cdot & \cdot \\ 0 & 0 & \cdots & 1 \end{vmatrix} = 1$$

this transformation of coordinates is a 1-1-mapping of a neighbourhood U of the origin of the (t,x)-space onto a neighbourhood U' of the origin of the (t',x)-space. Taking into consideration that $\lambda = \lambda(x)$ has a minimum at $x = 0$, this mapping into the (t',x)-space possesses the following properties:

It transforms all points of U with $t > 0$ onto a subset of U' whose intersection with hyperplanes $t' = \text{const}$ are compact subsets of the corresponding neighbourhood of the origin of the x-space (cf. the figure in section 5.2., step II; it may be added that the definition (5.32)

[1]) Other possible generalizations of the Holmgren theorem deal with distributional solutions instead of continuously differentiable solutions (cf. F. Treves' book [61]).

generalizes (5.8) and (5.27)).

Now introduce the new variables t', x instead of t, x into the given differential equation (5.3). In view of the chain rule we have

$$\frac{\partial u_k}{\partial x_i} = \frac{\partial \tilde{u}_k}{\partial t'}\frac{\partial \lambda}{\partial x_i} + \frac{\partial \tilde{u}_k}{\partial x_i}, \tag{5.33}$$

where $\tilde{u}$ denotes the given solution $u = u(t,x)$ in the new variables t', x, i.e.,

$$\tilde{u}(t',x) = u(t + \lambda(x), x). \tag{5.34}$$

Substituting (5.33) into (5.3) and solving the arising system for $\frac{\partial \tilde{u}_k}{\partial t'}$, we obtain a system which is similar to (5.3), namely

$$\frac{\partial \tilde{u}_j}{\partial t'} = \sum_{i=1}^{n}\sum_{k=1}^{m} \tilde{a}_{ik}^{(j)}(t',x)\frac{\partial \tilde{u}_k}{\partial x_i} + \sum_{k=1}^{m} \tilde{a}_k^{(j)}(t',x)\tilde{u}_k. \tag{5.35}$$

Since one may obtain this new system by applying Cramer's rule it is clear that the new coefficients $\tilde{a}_{ik}^{(j)}(t',x)$, $\tilde{a}_k^{(j)}(t',x)$ are rational functions in the derivatives $\partial\lambda/\partial x_i$ and in the former coefficients $a_{ik}^{(j)}(t' - \lambda(x), x)$, $a_k^{(j)}(t' - \lambda(x), x)$. In the special case of the system (5.5) the new coefficients are given by (5.10).

Now assume that the original coefficients are representable by power series in $x_1, \ldots, x_n$ whose coefficients are continuous functions depending on $t + \lambda(x)$, i.e., the original coefficients are given in the form

$$\sum_{\nu_1,\ldots,\nu_n} c_{\nu_1\ldots\nu_n}(t + \lambda(x))x_1^{\nu_1}\ldots x_n^{\nu_n}. \tag{5.36}$$

A function with such representation is called a <u>deformed power series</u> with respect to the point $x = 0$. Note that a deformed power series with respect to $x = 0$ is not necessarily also a deformed power series with respect to another point x_0 belonging to the domain of convergence.

It is immediately clear that the sum as well as the difference of two deformed power series is again a deformed power series. The same is also true for the product of two deformed power series because the product of two power series in $x_1, \ldots, x_n$ is again a power series in $x_1, \ldots, x_n$ and its coefficients are polynomials in the coefficients of the two given series. Finally, also the quotient of two deformed power series is a deformed power series. This can be proved in the following way:

Regard a series of type (5.36) with $c_{0\ldots0} \neq 0$. Then we look for a second series with coefficients $d_{\nu_1\ldots\nu_n}$ such that the product is everywhere equal to 1. Applying the identity theorem for power series to the product of the two series one obtains

$$c_{0\ldots 0}(t + \lambda(x))d_{0\ldots 0} = 1$$

$$c_{10\ldots 0}(t + \lambda(x))d_{0\ldots 0} + c_{0\ldots 0}(t + \lambda(x))d_{10\ldots 0} = 0$$

$$c_{010\ldots 0}(t + \lambda(x))d_{0\ldots 0} + c_{0\ldots 0}(t + \lambda(x))d_{010\ldots 0} = 0$$

and so on. Thus the desired coefficients $d_{\nu_1\ldots\nu_n}$ are rational functions in the given ones, $c_{\nu_1\ldots\nu_n}(t + \lambda(x))$ and they are, consequently, also continuous functions in $t + \lambda(x)$. This means that

$$\left(\sum_{\nu_1,\ldots,\nu_n}' c_{\nu_1\ldots\nu_n}(t + \lambda(x))x_1^{\nu_1}\ldots x_n^{\nu_n}\right)^{-1}$$

is a deformed power series, too, provided $c_{0\ldots 0}(t + \lambda(x)) \neq 0$ everywhere.

Summarizing these considerations, the following lemma has been proved:

<u>Lemma.</u> Sum, difference, product, and quotient of two deformed power series in $x_1, \ldots, x_n$ are again deformed power series.

Applying this lemma, it is immediately clear that the coefficients $\tilde{a}_{ik}^{(j)}(t',x)$, $\tilde{a}_k^{(j)}(t',x)$ of the transformed differential equations (5.35) are representable in the form

$$\sum_{\nu_1,\ldots,\nu_n} c_{\nu_1\ldots\nu_n}(t + \lambda(x))x_1^{\nu_1}\ldots x_n^{\nu_n}$$
$$= \sum_{\nu_1\ldots\nu_n} c_{\nu_1\ldots\nu_n}(t')x_1^{\nu_1}\ldots x_n^{\nu_n}$$

provided the coefficients of the originally given system (5.3) are deformed power series in $x_1, \ldots, x_n$.

Now let $u = u(t,x)$ be a given solution to (5.3) defined for $t \geq 0$ and vanishing identically at $t = 0$. First extend the given solution by setting

$$u(t,x) = 0 \quad \text{for} \quad t < 0$$

(cf. 5.2., I.). Next introduce (t',x) instead of (t,x) as new variable, where t' is given by (5.32). Then the function $\tilde{u}$ defined by (5.34) is a solution to the transformed equation (5.35) possessing a compact support for each t (cf. 5.2., II). If the coefficients of the originally given differential equation (5.3) are deformed power series with respect to $x = 0$, then the coefficients of the transformed equation (5.35) depend continuously on t. On the other hand, it is sufficient for completing the proof of the Holmgren theorem (5.2., III. - VI.) that the coefficients depend continuously on t: the coefficients in question do in fact not occur in the parts III. and VI. of the proof 5.2.; the parts IV. and V make use only of the continuity of the coefficients with respect to the variable t'. Thus the following theorem holds:

Theorem. Suppose that the coefficients of the system (5.3) of partial differential equations are deformed power series with respect to the point $x = 0$. Then there exists a neighbourhood of the origin of the (t,x)-space in which the solution is identically equal to zero provided its initial values are equal to zero everywhere.

Finally we would like to add that this generalized Holmgren theorem may be proved also by using the original method for proving the Holmgren theorem (cf. section 5.3.5.). In order to carry out the classical proof, we need a solution v to the differential equation (5.30) vanishing identically on the curve γ defined by $t = \psi(x)$. In 5.3.5. we constructed this solution under the assumption that the coefficients of the originally given differential equation are power series in both t and x. For the purpose of weakening this assumption we introduce new variables t', x instead of t, x by setting

$$t' = t - \psi(x).$$

Then the curve γ in the (t,x)-plane passes into an interval on the t'-axis in the (t',x)-plane, whereas the originally regarded interval [a, b] on the x-axis of the (t,x)-plane passes into a curve situated below the x-axis in the (t',x)-plane. Then transform the differential equation (5.30) into the (t',x)-plane. The coefficients of the transformed differential equation can be expressed by the coefficients of (5.30) and by the derivatives of $\psi = \psi(x)$. In order to ensure the existence of a solution of the transformed differential equation, we suppose that its coefficients are power series in t' and x. This results in the assumption that the original coefficients are deformed power-series.

6. BASIC PROPERTIES OF GENERALIZED ANALYTIC FUNCTIONS

Using the theory of initial value problems in scales of Banach spaces (cf. chapter 3), we constructed solutions to initial value problems with holomorphic initial functions in chapter 4. On the other hand, generalized analytic functions possess many common properties with holomorphic ones. Taking into accont these common properties, we intend to solve initial value problems with generalized analytic functions as initial functions in chapter 7. In the present chapter we collect basic definitions and basic properties of generalized analytic functions. The proofs of the statements and further details may be found in I. N. Vekua's monograph [77] and in L. Bers' lectures [7], see also [64]. In the present chapter proofs are carried out in detail only in such cases in which the statements are not formulated or are not proved completely in the quoted literatur. Concerning new tendencies of applications of complex analysis to partial differential equations see also the book [31] written by an international team of authors.

6.1. Partial complex differentiations in the classical sense and according to Sobolev

6.1.1. Let G be a given domain in the z-plane, $z = x + iy$. Assume, further, that $w = f(z)$ is a (complex-valued) continuously differentiable function defined in the domain G. Take any point $z_0 = x_0 + iy_0$ belonging to G. Then the linearization $\tilde{f}$ of f with respect to z_0 is defined by

$$\tilde{f}(z) = f(z_0) + c_1(x - x_0) + c_2(y - y_0), \qquad (6.1)$$

where

$$c_1 = \frac{\partial f}{\partial x}(x_0,y_0), \qquad c_2 = \frac{\partial f}{\partial y}(x_0,y_0).$$

Since

$$z - z_0 = (x - x_0) + i(y - y_0)$$

and

$$\overline{(z - z_0)} = (x - x_0) - i(y - y_0)$$

we have

$$x - x_0 = \frac{1}{2}\left[(z - z_0) + \overline{(z - z_0)}\right],$$
$$y - y_0 = \frac{i}{2}\left[(z - z_0) - \overline{(z - z_0)}\right].$$

Substituting these expressions into (6.1), we obtain

$$\tilde{f}(z) = f(z_0) + d_1(z - z_0) + d_2\overline{(z - z_0)}, \qquad (6.2)$$

where

$$d_1 = \frac{1}{2}\left[\frac{\partial f}{\partial x}(x_0,y_0) - i\frac{\partial f}{\partial y}(x_0,y_0)\right],$$
$$d_2 = \frac{1}{2}\left[\frac{\partial f}{\partial x}(x_0,y_0) + i\frac{\partial f}{\partial y}(x_0,y_0)\right].$$

The coefficients d_1 and d_2 in (6.2) are called the partial complex derivatives of f to z and $\bar{z}$ resp. at the point z_0 and are denoted by

$$\frac{\partial f}{\partial z}(z_0) \quad \text{and} \quad \frac{\partial f}{\partial \bar{z}}(z_0) \quad \text{resp.}$$

Thus the partial complex derivations are defined by

$$\frac{\partial f}{\partial z} = \frac{1}{2}\left(\frac{\partial f}{\partial x} - i\frac{\partial f}{\partial y}\right), \qquad (6.3)$$

$$\frac{\partial f}{\partial \bar{z}} = \frac{1}{2}\left(\frac{\partial f}{\partial x} + i\frac{\partial f}{\partial y}\right). \qquad (6.4)$$

The last two formulae allow us, further, to express the derivatives $\frac{\partial f}{\partial x}$ and $\frac{\partial f}{\partial y}$ by $\frac{\partial f}{\partial z}$ and $\frac{\partial f}{\partial \bar{z}}$:

$$\frac{\partial f}{\partial x} = \frac{\partial f}{\partial z} + \frac{\partial f}{\partial \bar{z}}, \quad \frac{\partial f}{\partial y} = i\left(\frac{\partial f}{\partial z} - \frac{\partial f}{\partial \bar{z}}\right).$$

6.1.2. The partial complex differentiations $\frac{\partial}{\partial z}$ and $\frac{\partial}{\partial \bar{z}}$ defined in 6.1.1. satisfy the well-known rules about the derivative of linear combinations, products and so on, too. For example, the derivative of the product $f_1 f_2$ with respect to $\bar{z}$ is given by

$$\frac{\partial}{\partial \bar{z}}(f_1 f_2) = f_1\frac{\partial f_2}{\partial \bar{z}} + f_2\frac{\partial f_1}{\partial \bar{z}}.$$

The composition $g \circ f$ of two differentiable functions $w = f(z)$ and $W = g(w)$ is differentiable, too, and the chain rule may be written in the form

$$\frac{\partial}{\partial z}(g \circ f) = \frac{\partial g}{\partial w}\frac{\partial w}{\partial z} + \frac{\partial g}{\partial \bar{w}}\frac{\partial \bar{w}}{\partial z} \quad \text{and}$$

$$\frac{\partial}{\partial \bar{z}}(g \circ f) = \frac{\partial g}{\partial w}\frac{\partial w}{\partial \bar{z}} + \frac{\partial g}{\partial \bar{w}}\frac{\partial \bar{w}}{\partial \bar{z}}.$$

Note, finally, that

$$\frac{\partial \bar{w}}{\partial z} = \overline{\left(\frac{\partial w}{\partial \bar{z}}\right)}, \quad \frac{\partial \bar{w}}{\partial \bar{z}} = \overline{\left(\frac{\partial w}{\partial z}\right)} \quad \text{and so on.}$$

6.1.3. Provided the (real- or complex-valued) function h is continuously differentiable in the closure of G we have

$$\iint_G \frac{\partial h}{\partial x}dxdy = \int_{\partial G} hdy \qquad (6.5)$$

$$\iint_G \frac{\partial h}{\partial y}dxdy = -\int_{\partial G} hdx, \qquad (6.6)$$

where ∂G denotes the (smooth) boundary of G. The formulae (6.5) and (6.6) resp. are the well-known Ostrogradski-Gauss integral formulae in the case of the plane. In view of (6.4) and $dy - idx = -i(dx + idy) = -idz$ one obtains

$$\iint_G \frac{\partial h}{\partial \bar{z}}dxdy = \frac{1}{2i}\int_{\partial G} hdz \qquad (6.7)$$

if to the equation (6.5) the equation (6.6) multiplied by i is to be added. Analogously, we obtain

$$\iint_G \frac{\partial h}{\partial z}dxdy = -\frac{1}{2i}\int_{\partial G} hd\bar{z} \qquad (6.8)$$

by subtracting the equation (6.6) after multiplication with i from the equation (6.5) if one takes into consideration the definition (6.3) and the relation $dy + idx = i(dx - idy) = id\bar{z}$.

6.1.4. Let f be a given continuously differentiable function defined in the domain G of the z-plane. Let, further, ϕ be any test function, i.e., ϕ is a continuously differentiable function in G which is identically equal to zero outside a compact subset of G. Thus ϕ and the product $f\phi$ vanish everywhere in a neighbourhood of the boundary of G. Denote the product $f\phi$ by h. Applying the complex form (6.7) of the Ostrogradski-Gauss integral formula to h, we obtain, therefore,

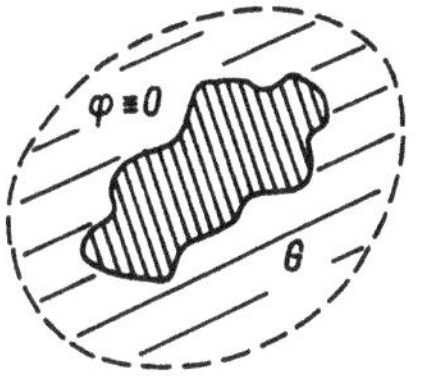

$$\iint_G \frac{\partial}{\partial \bar{z}}(f\phi)\,dxdy = 0.$$

Carrying out the differentiation of the integrand by using the product rule (see 6.1.2.), we get

$$\iint_G \left(f\frac{\partial \phi}{\partial \bar{z}} + \frac{\partial f}{\partial \bar{z}}\phi\right)dxdy = 0 \qquad (6.9)$$

for each test function ϕ. Denote the derivative $\partial f/\partial \bar{z}$ by g. Hence we have

$$\iint_G \left(f\frac{\partial \phi}{\partial \bar{z}} + g\phi\right)dxdy = 0 \qquad (6.10)$$

for each test function ϕ if g is the derivative of f to $\bar{z}$.

Now assume that f is any (integrable) function not assumed to be continuously differentiable. If there exists an (integrable) function g such that relation (6.10) is satisfied for each (continuously differentiable) test function ϕ the function g is said to be the derivative of f with respect to $\bar{z}$ in Sobolev's sense and is denoted by $\partial f/\partial \bar{z}$ like the classical derivative.

By comparison of (6.9) and (6.10) we see immediately that the classical derivative $\partial f/\partial \bar{z}$ in the case of its existence may be interpreted as derivative in Solbolev's sense, too.

Starting from the Ostrogradski-Gauss formula (6.8), one defines the derivative $g = \partial f/\partial z$ in Sobolev's sense analogously by the relation

$$\iint_G \left(f\frac{\partial \phi}{\partial z} + g\phi\right)dxdy = 0$$

which must be satisfied by every test function ϕ.

6.1.5. Suppose that $f = u + iv$ is holomorphic. Then the real part u and the imaginary part v of f satisfy the well-known Cauchy-Riemann system

$$\frac{\partial u}{\partial x} = \frac{\partial v}{\partial y}, \quad \frac{\partial v}{\partial x} = -\frac{\partial u}{\partial y}. \qquad (6.11)$$

Thus holomorphic functions satisfy the relation

$$\frac{\partial f}{\partial x} = \frac{\partial u}{\partial x} + i\frac{\partial v}{\partial x} = \frac{\partial v}{\partial y} - i\frac{\partial u}{\partial y} = - i\left(\frac{\partial u}{\partial y} + i\frac{\partial v}{\partial y}\right) = - i\frac{\partial f}{\partial y}. \tag{6.12}$$

In view of (6.4) we get, consequently, the relation

$$\frac{\partial f}{\partial \bar{z}} = 0 \tag{6.13}$$

for every holomorphic function f. It may be added that the last equation (6.13) is nothing else than the complex form of the Cauchy-Riemann system (6.11).

Further it is well-known that the value of the complex derivative

$$\frac{df}{dz}(z_0) = \lim_{z \to z_0} \frac{f(z) - f(z_0)}{z - z_0}$$

of a holomorphic function f is equal to the value of the partial derivative with respect to x, i.e., we have

$$\frac{df}{dz}(z_0) = \frac{\partial f}{\partial x}(z_0). \tag{6.14}$$

On the other hand, by virtue of (6.12) the partial complex derivative $\partial f/\partial z$ defined by (6.3) equals to

$$\frac{\partial f}{\partial z} = \frac{\partial f}{\partial x}$$

in the case of a holomorphic function f. Comparing this expression with (6.14), we obtain, finally, that the ordinary complex derivative df/dz and the partial complex derivative $\partial f/\partial z$ coincide for holomorphic f.

6.1.6. Let f be an arbitrary holomorphic function in G, i.e., the relation (6.13) holds. In view of (6.10) the function f satisfies, therefore, the relation

$$\iint_G f\frac{\partial \phi}{\partial \bar{z}}dxdy = 0 \tag{6.15}$$

for every test function ϕ.

Conversely assume now that f is any (integrable) function satisfying the relation (6.15) for any test function ϕ. Then the famous Weyl lemma states that f is necessarily a holomorphic function in the ordinary classical sense. Provided f is continuous, an elementary proof of Weyl's lemma is given in [64], for instance.

6.1.7. It may be added that formula (6.7) yields Cauchy's integral theorem, too: if h is holomorphic in G, then in view of (6.13) we have $\partial h/\partial \bar{z} = 0$ everywhere in G and, consequently, the right-hand side of (6.67) equals to zero. Therefore one gets

$$\int_{\partial G} h dz = 0.$$

6.1.8. Assume that f depends holomorphically on n complex variables z_1, ..., z_n. Then in view of 6.1.5. the function f satisfies the system

$$\frac{\partial f}{\partial \bar{z}_i} = 0, \quad j = 1, \ldots, n.$$

Moreover, the ordinary complex derivatives with respect to z_j may be interpreted as partial complex derivatives $\partial f/\partial z_j$ in the sense of definition (6.3), i.e., we have

$$\frac{\partial f}{\partial z_j} = \frac{1}{2}\left(\frac{\partial f}{\partial x_j} - i\frac{\partial f}{\partial y_j}\right),$$

where $z_j = x_j + iy_j$.

6.2. Complex integral operators connected with the partial complex differentiations

6.2.1. Let G be a bounded domain in the z-plane and let, further, h be an integrable function defined in G. Define a new function $H = H(z)$ by

$$H(z) = -\frac{1}{\pi}\iint_G \frac{h(\zeta)}{\zeta - z}d\xi\, d\eta, \tag{6.16}$$

where $\zeta = \xi + i\eta$ and z is an arbitrary point in the z-plane. Introducing polar coordinates with the point z as centre, the singularity $1/(\zeta - z)$ can be eliminated. Thus the integral exists in the ordinary sense even in the case that z belongs to $\bar{G}$.

For short we denote the function $H = H(z)$, that depends on h, by $T_G h$. Thus T_G is an integral operator transforming functions defined in G into such ones being defined in the whole z-plane. Later on we investigate the restriction of $T_G h$ to G and $\bar{G}$ again denoted by $T_G h$.

6.2.2. Since the integrand of the integral (6.16) depends holomorphically on z (outside $\bar{G}$) the function $H = H(z)$ is proved to be holomorphic outside $\bar{G}$. Using Fubini's theorem, it is easy to prove that in G the derivative of $T_G h$ with respect to $\bar{z}$ in Sobolev's sense is equal to h, i.e., $T_G h$ satisfies the differential equation

$$\frac{\partial}{\partial \bar{z}}T_G h = h \tag{6.17}$$

in G. In other words, the differential operator $\frac{\partial}{\partial \bar{z}}$ is proved to be left-inverse to T_G.

Let us additionally remark that the operator $\bar{T}_G$ defined by

$$(\bar{T}_G h)[z] = -\frac{1}{\pi}\iint_G \frac{h(\zeta)}{\bar{\zeta} - \bar{z}}d\xi\, d\eta$$

satisfies the differential equation

$$\frac{\partial}{\partial z}\bar{T}_G h = h,$$

analogously.

6.2.3. In view of (6.17) the function $T_G h$ is a special solution to the so-called inhomogeneous Cauchy-Riemann equation

$$\frac{\partial w}{\partial \bar{z}} = h \tag{6.18}$$

in G, where h is given. By virtue of Weyl's lemma 6.1.6. the general solution of the differential equation (6.18) is given by

$$w = T_G h + \Phi,$$

where Φ is an arbitrary holomorphic function.

6.2.4. A function h is said to be Hölder-continuous with the exponent (or index) α, $0 < \alpha \leq 1$, if there exists a constant C such that

$$|h(z_1) - h(z_2)| \leq C|z_1 - z_2|^\alpha \tag{6.19}$$

for any two points z_1, z_2 belonging to the set in which h is defined. Notice that the constant C may depend on the choice of the function h. In case that (6.19) is satisfied with $\alpha = 1$, the function is said to be Lipschitz-continuous.

6.2.5. Replacing $|z_2 - z_1|$ by the distance of two points, definition 6.2.4. of the Hölder-continuity can be generalized to the case of mappings between metric spaces.

6.2.6. Let M be a bounded subset of the z-plane. The space of all (complex-valued) functions defined in M and Hölder-continuous with the exponent α, $0 < \alpha \leq 1$, is denoted by $C^\alpha(M)$. This space equipped with the norm

$$\|h\| = \max\left\{\sup_M |h(z)|,\ \sup_{z_1 \neq z_2} \frac{|h(z_2) - h(z_1)|}{|z_2 - z_1|^\alpha}\right\}$$

turns out ot be a Banach space.

The norm $\|h\|$ defined above is also denoted by $\|h\|_{C^\alpha(M)}$ if we want to express that the Hölder-continuous function h is defined in the set M. Such a denotation is necessary if simultaneously Hölder-continuous functions defined in sets different from each other are investigated (in section 6.2.11., for instance, we compare norms in $C^\alpha(\bar{G}')$ with those in $C^\alpha(\bar{G})$, where G' is a subdomain of G).

6.2.7. The T_G-operator defined in 6.2.1. is a linear and bounded operator mapping $C^\alpha(\bar{G})$ into itself, where G is a given bounded domain in the z-plane.

6.2.8. Let G be any open disk in the z-plane and K any closed disk contained in G. Then each (complex-valued) function h continuously differentiable in G turns out to be Hölder-continuous with the exponent α in K, where α is chosen arbitrarily, $0 < \alpha \leq 1$. The proof of this statement is based on the mean value theorem of the differential calculus:

First assume that h is real-valued. Then for any two points z_1 and z_2 belonging to K we have

$$h(z_2) - h(z_1) = \frac{\partial h}{\partial x}(\tilde{z})(x_2 - x_1) + \frac{\partial h}{\partial y}(\tilde{z})(y_2 - y_1),$$

where $z_j = x_j + iy_j$, $j = 1, 2$, and $\tilde{z}$ is a suitably chosen point lying between z_1, z_2 on the straight line through these two points. Since $|x_2 - x_1|, |y_2 - y_1| \leq |z_2 - z_1|$ we get

$$|h(z_2) - h(z_1)| \leq 2c|z_2 - z_1|,$$

where c is an upper bound of the modules of the first order derivatives of h in K. From the last inequality one obtains the desired estimate (6.19) with

$$C = 2cd^{1-\alpha},$$

where d is the diameter of K.

In the case of a complex-valued function h one applies the latter consideration to both the real part and the imaginary part of h.

6.2.9. Let G be any open subset of the z-plane. Moreover suppose that the function h is defined and continuously differentiable in G. Let, further, K be any compact subset of G. Then the following statement is true:

h is Hölder-continuous with the exponent α in K, where α is any number satisfying the inequality $0 < \alpha \leq 1$.

Proof. Denote the distance of K from the boundary ∂G of G by δ, $\delta > 0$. Take any point z_0 belonging to K. Then the open disk $U_\delta(z_0)$ centred at z_0 with radius δ is contained in G. Thus the given function h is defined and continuously differentiable in the whole open disk $U_\delta(z_0)$.

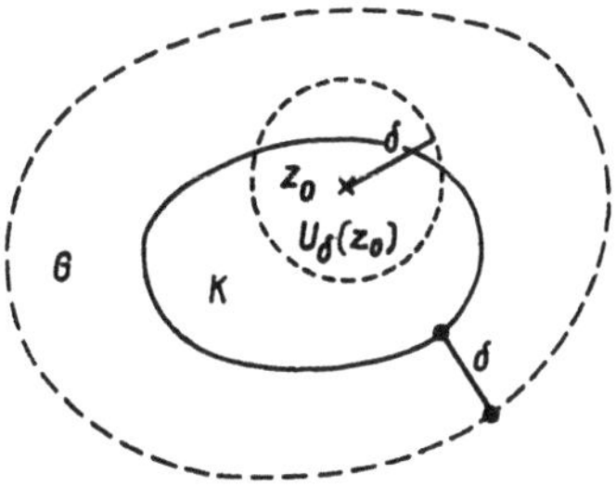

Now take into consideration that the set of all open disks $U_{\frac{1}{3}\delta}(z_0)$ with the radii $\frac{1}{3}\delta$ form an open covering of K if z_0 varies in K. Therefore K is contained in the union of a finite number of open disks $U_{\frac{1}{3}\delta}(z_{oi})$, where the points z_{oi}, $i = 1, \ldots, m$, are contained in K. Since the closed disk $\overline{U_{\frac{2}{3}\delta}(z_{oi})}$ centred at z_{oi} with radii $\frac{2}{3}\delta$ are subsets of the open disks $U_\delta(z_{oi})$, in view of 6.2.8. the given function h is Hölder-continuous with exponent α in $\overline{U_{\frac{2}{3}\delta}(z_{oi})}$, especially, i.e., the estimate (6.19) with a certain constant C_i instead of C, $i = 1, \ldots, m$, is true in those closed disks.

In order to conclude the proof of the above statement, we take any two points z_1 and z_2 belonging to K. We have to show that the estimate (6.19) holds with a suitably chosen constant C depending on K but not depending on z_1, z_2. Firstly, assume that the inequality

$$|z_2 - z_1| \leq \frac{1}{3}\delta \tag{6.20}$$

holds. The point z_1 is contained at least in one open disk $U_{\frac{1}{3}\delta}(z_{oj})$ with a suitably chosen j, $1 \leq j \leq m$, since the open disks $U_{\frac{1}{3}\delta}(z_{oi})$, $i = 1, \ldots, m$, form a covering of K. By virtue of (6.20) the point z_2 is contained in $U_{\frac{2}{3}\delta}(z_{oj})$. Hence (6.19) holds with C_j instead of C.

Secondly, assume that in distinction from (6.20) the two given points satisfy the inequality

$$|z_2 - z_1| > \frac{1}{3}\delta.$$

Then one has

$$1 < \frac{3}{\delta}|z_2 - z_1| = \frac{3}{\delta}|z_2 - z_1|^{1-\alpha}|z_2 - z_1|^{\alpha}. \tag{6.21}$$

On the other hand we have

$$|h(z_2) - h(z_1)| \leq 2 \sup_K |h|.$$

Replacing the factor 2 by the right-hand side of (6.21) multiplied by 2, we see that (6.19) holds also in the second case if C is replaced by

$$\frac{6}{\delta} \sup_K |h| \cdot d^{1-\alpha}, \tag{6.22}$$

where d is the diameter of K.

Summarizing these estimates, it follows that (6.19) is true in any case in K, where C is equal to the maximum of $C_1, \ldots, C_m$ and the quantity (6.22).

Note, finally, that continuously differentiable functions defined in open sets in R^n are Hölder-continuous in compact subsets, too (concerning the concept of Hölder-continuity cf. 6.2.5.).

6.2.10. Now we are going to show that a function is not necessarily Hölder-continuous with exponent α in the whole domain of a function if it is Hölder-continuous with exponent α in each compact subset of the domain of the function. To this end we regard the interval (0, 1) on the x-axis. Regard the function h defined by

$$h(x) = x^{\beta},$$

where β is a fixed real number, $0 < \beta < 1$. Let α be a further fixed number, $\beta < \alpha \leqq 1$. Take any two points x_1, x_2 belonging to the interval $(0, 1)$. Applying the mean value theorem of the differential calculus, one obtains

$$h(x_2) - h(x_1) = \beta\tilde{x}^{\beta-1}(x_2 - x_1), \tag{6.23}$$

where $\tilde{x}$ is suitably chosen between x_1 and x_2. Assume, especially, that

$$x_1 = c, \quad x_2 = 2c,$$

where $0 < c < \frac{1}{2}$. Then we have $x_2 - x_1 = c$ and, consequently,

$$|x_2 - x_1| = |x_2 - x_1|^{1-\alpha}|x_2 - x_1|^{\alpha} = c^{1-\alpha}|x_2 - x_1|^{\alpha}. \tag{6.24}$$

On the other hand in view of $c < \tilde{x} < 2c$ one obtains

$$\frac{1}{\tilde{x}} > \frac{1}{2c}$$

and, therefore,

$$\tilde{x}^{\beta-1} > \frac{1}{2^{1-\beta}c^{1-\beta}}\,. \tag{6.25}$$

Taking into account (6.24) and (6.25), the representation (6.23) of $h(x_2) - h(x_1)$ yields the estimate

$$|h(x_2) - h(x_1)| > \frac{\beta}{2^{1-\beta}}c^{\beta-\alpha}|x_2 - x_1|^{\alpha}.$$

Since $\beta - \alpha < 0$ the coefficient of $|x_2 - x_1|^{\alpha}$ is arbitrarily large if c is sufficiently small. Hence the function h in question is not Hölder-continuous with the exponent α, $\beta < \alpha \leqq 1$, in the whole interval $(0, 1)$ although h is continuously differentiable in $(0, 1)$ and, consequently, Hölder-continuous in each compact subset of $(0, 1)$ (cf. 6.2.9.).

<u>6.2.11.</u> Let G be a bounded domain in the z-plane. Let further Φ be holomorphic in G and Hölder-continuous in $\overline{G}$ with the exponent α. The holomorphy of Φ implies, especially, that Φ is continuously differentiable in G.

Moreover suppose that G' is a subdomain of G whose closure $\overline{G'}$ is a compact subset of G. In view of 6.2.9. the derivative Φ' is Hölder-continuous in $\overline{G'}$, in particular. Denote the distance of G' from the boundary ∂G of G by δ.

Then the following estimate connecting the norms $\|\Phi'\|_{C^{\alpha}(\overline{G'})}$ and $\|\Phi\|_{C^{\alpha}(\overline{G})}$ (defined in section 6.2.6.) holds:

$$\|\Phi'\|_{C^{\alpha}(\overline{G'})} \leqq \frac{3\cdot 2^{\alpha}}{\delta}\|\Phi\|_{C^{\alpha}(\overline{G})}\,. \tag{6.26}$$

<u>Proof.</u> Let z be any point belonging to $\overline{G'}$. Then the disk centred at z

with radius δ is contained in $\overline{G}$. In view of Cauchy's integral formula we have

$$\Phi'(z) = \frac{1}{2\pi i}\int_{|\zeta-z|=\delta} \frac{\Phi(\zeta)}{(\zeta - z)^2}\,d\zeta. \tag{6.27}$$

The definition 6.2.6. of $\|\Phi\|_{C^\alpha(\overline{G})}$ implies that

$$|\Phi(\zeta)| \leq \|\Phi\|_{C^\alpha(\overline{G})}$$

for each ζ in question. Thus (6.27) yields

$$|\Phi'(z)| \leq \frac{1}{2\pi}\,\frac{1}{\delta^2}\|\Phi\|_{C^\alpha(\overline{G})}\cdot 2\pi\delta$$

and, therefore,

$$\sup_{G'}|\Phi'(z)| \leq \frac{1}{\delta}\|\Phi\|_{C^\alpha(\overline{G})}. \tag{6.28}$$

Taking into account the relation

$$\int_{|\zeta-z|=\delta} \frac{1}{(\zeta - z)^2}\,d\zeta = 0, \tag{6.29}$$

the formula (6.27) may be rewritten as

$$\Phi'(z) = \frac{1}{2\pi i}\int_{|\zeta-z|=\delta} \frac{\Phi(\zeta) - \Phi(z)}{(\zeta - z)^2}\,d\zeta. \tag{6.30}$$

On the other hand, by virtue of 6.2.6. the module of $\Phi(\zeta) - \Phi(z)$ may be estimated by

$$|\Phi(\zeta) - \Phi(z)| \leq \|\Phi\|_{C^\alpha(\overline{G})}|\zeta - z|^\alpha.$$

Therefore formula (6.30) leads to

$$|\Phi'(z)| \leq \frac{1}{2\pi}\,\frac{1}{\delta^{2-\alpha}}\|\Phi\|_{C^\alpha(\overline{G})}\cdot 2\pi\delta$$

and, consequently, to the estimate

$$\sup_{G'}|\Phi'(z)| \leq \frac{1}{\delta^{1-\alpha}}\|\Phi\|_{C^\alpha(\overline{G})}. \tag{6.31}$$

Now take any two points z_1, z_2 belonging to $\overline{G'}$. Since the disk centred at z_1 with radius δ belongs to $\overline{G}$ formula (6.27) holds with z_1 instead of z. Analogously we have

$$\Phi'(z_2) = \frac{1}{2\pi i}\int_{|\zeta-z_1|=\delta} \frac{\Phi(\zeta)}{(\zeta - z_2)^2}\,d\zeta$$

provided z_2 satisfies the inequality

$$|z_2 - z_1| < \delta \tag{6.32}$$

implying that z_2 belongs to the open disk centred at z_1 with radius δ. Summarizing these representation formulae, one obtains

$$\Phi'(z_2) - \Phi'(z_1) = \frac{1}{2\pi i}\int_{|\zeta-z_1|=\delta} \Phi(\zeta)\left(\frac{1}{(\zeta-z_2)^2} - \frac{1}{(\zeta-z_1)^2}\right)d\zeta. \tag{6.33}$$

The relation (6.29) holds with z_1 instead of z, too. Similarly we have

$$\int_{|\zeta-z_1|=\delta} \frac{1}{(\zeta - z_2)^2} d\zeta = 0.$$

Hence formula (6.33) may be rewritten as

$$\Phi'(z_2) - \Phi'(z_1) = \frac{1}{2\pi i} \int_{|\zeta-z_1|=\delta} (\Phi(\zeta) - \Phi(z_1)) \left(\frac{1}{(\zeta - z_2)^2} - \frac{1}{(\zeta - z_1)^2}\right) d\zeta. \tag{6.34}$$

Once more taking into consideration the definition 6.2.6. of the norm, one gets the estimate

$$|\Phi(\zeta) - \Phi(z_1)| \leq \|\Phi\|_{C^\alpha(\overline{G})} |\zeta - z_1|^\alpha.$$

On the other hand we have

$$\frac{1}{(\zeta - z_2)^2} - \frac{1}{(\zeta - z_1)^2} = \frac{(z_2 - z_1)(2\zeta - z_1 - z_2)}{(\zeta - z_1)^2(\zeta - z_2)^2}$$

$$= (z_2 - z_1)\left(\frac{1}{(\zeta - z_1)(\zeta - z_2)^2} + \frac{1}{(\zeta - z_1)^2(\zeta - z_2)}\right).$$

Thus the module of the integrand of (6.34) may be estimated by

$$|z_2 - z_1| \left(\frac{1}{|\zeta - z_1|^{1-\alpha}|\zeta - z_2|^2} + \frac{1}{|\zeta - z_1|^{2-\alpha}|\zeta - z_2|}\right) \|\Phi\|_{C^\alpha(\overline{G})}.$$

Strengthening the assumption (6.32), first we regard pairs z_1, z_2 satisfying the inequality

$$|z_2 - z_1| \leq \frac{\delta}{2}. \tag{6.35}$$

Then in view of $|\zeta - z_1| = \delta$ we have $|\zeta - z_2| \geq \frac{\delta}{2}$ and, therefore,

$$\frac{1}{|\zeta - z_2|} \leq \frac{2}{\delta}.$$

Thus the representation formula (6.34) leads to the estimate

$$|\Phi'(z_2) - \Phi'(z_1)|$$
$$\leq \frac{1}{2\pi}|z_2 - z_1| \left(\frac{1}{\delta^{1-\alpha}}\left(\frac{2}{\delta}\right)^2 + \frac{1}{\delta^{2-\alpha}}\,\frac{2}{\delta}\right) \|\Phi\|_{C^\alpha(\overline{G})} \cdot 2\pi\delta$$
$$\leq \frac{6}{\delta^{2-\alpha}}|z_2 - z_1| \cdot \|\Phi\|_{C^\alpha(\overline{G})}.$$

In view of (6.35) we have, moreover,

$$|z_2 - z_1| = |z_2 - z_1|^{1-\alpha}|z_2 - z_1|^\alpha \leq \left(\frac{\delta}{2}\right)^{1-\alpha} |z_2 - z_1|^\alpha$$

and, consequently,

$$|\Phi'(z_2) - \Phi'(z_1)| \leq \frac{3 \cdot 2^\alpha}{\delta} \|\Phi\|_{C^\alpha(\overline{G})} |z_2 - z_1|^\alpha. \tag{6.36}$$

Finally regard points z_1, z_2 satisfying the inequality

$$|z_2 - z_1| > \frac{\delta}{2} \tag{6.37}$$

opposite to (6.35). The last inequality is equivalent to

$$1 < \frac{2}{\delta}|z_2 - z_1|.$$

In the case (6.37) we have, therefore,

$$1 < \left(\frac{2}{\delta}\right)^{\alpha}|z_2 - z_1|^{\alpha}, \tag{6.38}$$

too. On the other hand the inequality (6.31) implies

$$|\Phi'(z_2) - \Phi'(z_1)| \leqq 2 \sup_{G'}|\Phi'(z)| \leqq \frac{2}{\delta^{1-\alpha}} \|\Phi\|_{C^\alpha(\overline{G})} \cdot 1.$$

Replacing the factor 1 by the larger right-hand side of (6.38), we obtain the estimate

$$|\Phi'(z_2) - \Phi'(z_1)| < \frac{2\cdot 2^{\alpha}}{\delta}\|\Phi\|_{C^\alpha(\overline{G})}\cdot|z_2 - z_1|^{\alpha} \tag{6.39}$$

that holds under the assumption (6.37).

The estimates (6.28), (6.36), and (6.39) show that our statement (6.36) is true.

In the end we would like to remark that in a similar way it is also possible to estimate the Hölder-norm 6.2.6. of higher derivatives of a holomorphic function. In the case of the n-th derivative the factor $\frac{1}{\delta}$ in (6.26) is to be replaced by $\frac{1}{\delta^n}$ (see [36]).

<u>6.2.12.</u> In addition to the T_G-operator, introduced by the right-hand side of formula (6.16) in section 6.2.1., we need the so-called Π_G-operator that is defined by

$$(\Pi_G h)[z] = -\frac{1}{\pi}\iint_G \frac{h(\zeta)}{(\zeta - z)^2}d\xi\, d\eta, \tag{6.40}$$

i.e., the weak singularity $\frac{1}{\zeta - z}$ in (6.16) is replaced by the strong one $\frac{1}{(\zeta - z)^2}$. In this case it is not possible to interpret the integral as an ordinary one by introducing polar coordinates, as we did in the case of the T_G-operator in section 6.2.1. The integral can be defined, however, as <u>Cauchy's principal value</u>, i.e., first we omit a δ-neighbourhood of z by the integration and define the integral as limit of the approximate integrals by the limiting process $\delta \to 0$. Moreover, the right-hand side of (6.40) may be rewritten as

$$-\frac{h(z)}{\pi}\iint_G \frac{1}{(\zeta - z)^2}d\xi\, d\eta - \frac{1}{\pi}\iint_G \frac{h(\zeta) - h(z)}{(\zeta - z)^2}d\xi\, d\eta. \tag{6.41}$$

Using the Ostrogradski-Gauss integral formula (cf. 6.1.3.), the Cauchy principal value of the first term can be represented by

$$-\frac{h(z)}{2\pi i}\int_{\partial G}\frac{\overline{\zeta}}{(\zeta - z)^2}\,d\zeta \tag{6.42}$$

provided z is an (interior) point of G. Additionally assume now that h is Hölder-continuous with exponent α, $0 < \alpha < 1$ (cf. 6.2.4.). Then the integrand of the second integral in (6.41) may be estimated by

$$\frac{|h(\zeta) - h(z)|}{(\zeta - z)^2} \leqq \frac{C|\zeta - z|^\alpha}{|\zeta - z|^2} = \frac{C}{|\zeta - z|^{2-\alpha}} .$$

Introducing polar coordinates with the centre z, the singularity of the integrand can be reduced to

$$\frac{1}{|\zeta - z|^{1-\alpha}} .$$

Denoting $|\zeta - z|$ by r and taking into account the formula

$$\int \frac{1}{r^{1-\alpha}}\, dr = r^\alpha, \tag{6.43}$$

we see, consequently, that the second integral in (6.41) exists as ordinary integral.

Summarizing the above arguments, one obtains the result that $\Pi_G h$ may be represented by the ordinary integrals (6.42) and (6.43) resp. provided the given function h is Hölder-continuous.

Finally it should be remarked that $\Pi_G h$ may be defined for functions h belonging to $L_p(G)$, $p > 1$, too. Approximating such functions h by Hölder-continuous ones, h_n, the singular integral $\Pi_G h$ is defined as the limit of $\Pi_G h_n$ (cf. also 6.2.15.).

6.2.13. There exists an important connection beween the T_G- and the Π_G-operators, introduced in 6.2.1. and 6.2.12., resp. We formulate it in the case of a bounded domain G and an integrand h Hölder-continuous in $\overline{G}$. In view of (6.17) the derivative of $T_G h$ with respect to $\overline{z}$ equals to h. Using the concept of derivatives in Sobolev's sense (cf. 6.1.4.), applying Fubini's theorem, and taking into consideration the relation

$$\frac{\partial}{\partial z}\left(\frac{1}{\zeta - z}\right) = \frac{1}{(\zeta - z)^2} ,$$

one is able to prove that the derivative of $T_G h$ with respect to z is equal to $\Pi_G h$. Thus in addition to (6.17) the formula

$$\frac{\partial}{\partial z} T_G h = \Pi_G h \tag{6.44}$$

is true. The formulae (6.17) and (6.44) also allow us to express the derivatives of $T_G h$ with respect to x and y by h and $\Pi_G h$ because the derivatives with respect to real variables may be expressed by those with respect to z and $\overline{z}$ (cf. 6.1.1.).

6.2.14. In addition to the statement in 6.2.7. the following is true: The Π_G-operator is a linear and bounded operator mapping $C^\alpha(\overline{G})$ into itself, where $0 < \alpha < 1$ and G is a bounded domain in the z-plane.

6.2.15. To conclude we would like to make mention of some further properties of the T_G- and the Π_G-operators, where G is assumed to be bounded.

First mention that the T_G-operator is a linear and bounded operator mapping $C^\alpha(\overline{G})$ into $C^{1+\alpha}(\overline{G})$, where $C^{1+\alpha}(\overline{G})$ denotes the space of functions which are Hölder-contiuuously differentiable in the closure $\overline{G}$, $0 < \alpha < 1$.

Next notice that T_G maps $L_p(G)$ into $C^\beta(\overline{G})$ if $p > 2$ and $\beta = \frac{p-2}{p}$.

Finally, the T_G- and the Π_G-operators are linear and bounded operators mapping $L_p(G)$ into itself if $p > 1$. Remember that the L_p-norm of h is defined by

$$\|h\|_{L_p(G)} = \left(\iint_G |h|^p dxdy\right)^{\frac{1}{p}}. \tag{6.45}$$

6.2.16. Again let G be a bounded domain in the z-plane. Let further G' be any subdomain whose closure $\overline{G'}$ is compact in G. Denote the distance of G' from the boundary ∂G of G by δ.

Let Φ be a holomorphic function in G whose L_p-norm $\|\Phi\|_{L_p(G)}$ defined by (6.45) is finite, where $p > 1$.

Then A. Crodel [15] proved that the L_p-norm of Φ' in G' may be estimated by

$$\|\Phi'\|_{L_p(G')} \leq \frac{1}{\delta}\|\Phi\|_{L_p(G)}. \tag{6.46}$$

This estimate is analogous to the estimate (6.26) which is valid for the norm of holomorphic functions in the C^α-spaces of Hölder-continuous functions.

The proof of formula (6.46) is based on Fubini's theorem. In order to prove (6.46), take any point z belonging to G'. Then the closed disk centred at z with radius r is contained in G provided $r < \delta$. In view of Cauchy's integral formula (6.27) we have

$$\Phi'(z) = \frac{1}{2\pi i}\int_{|\zeta'|=r} \frac{\Phi(z+\zeta')}{\zeta'^2}d\zeta',$$

where $z - \zeta$ is denoted by ζ'. Applying Hölder's inequality, one obtains

$$|\Phi'(z)| \leq \frac{1}{2\pi}\left(\int_{|\zeta'|=r} |\Phi(z+\zeta')|^p|d\zeta'|\right)^{\frac{1}{p}}\left(\int_{|\zeta'|=r}\frac{1}{|\zeta'|^{2q}}|d\zeta'|\right)^{\frac{1}{q}},$$

where q is the exponent conjugate to p, i.e.,

$$\frac{1}{p} + \frac{1}{q} = 1.$$

The last estimate yields immediately

$$|\Phi'(z)|^p \leq \frac{1}{2\pi r^{p+1}}\int_{|\zeta'|=r} |\Phi(z+\zeta')|^p|d\zeta'|$$

and

$$\iint_{G'} |\Phi'|^p dxdy \leq \frac{1}{2\pi r^{p+1}} \int_{|\zeta'|=r} \left(\iint_{G'} |\Phi(z+\zeta')|^p dxdy \right) |d\zeta'|$$

after applying the Fubini theorem as mentioned above, where $z = x + iy$. Taking into account that the set of all points of the form $z + \zeta'$, where z varies arbitrarily in G' is a subset of G and carrying out the limiting process $r \to \delta$, the asserted estimate (6.46) follows immediately.

6.3. Generalized analytic functions

6.3.1. Let G be a given domain in the z-plane and $a = a(z)$, $b = b(z)$ two given functions defined and continuous in G. Then every solution $w = w(z)$ to the differential equation

$$\frac{\partial w}{\partial \bar{z}} = a(z)w + b(z)\bar{w} \tag{6.47}$$

is called a generalized analytic function. Remark that the derivative $\partial w/\partial \bar{z}$ may be understood in Sobolev's sense (cf. 6.1.4.). To simplify matters we assume that w as well as $\partial w/\partial \bar{z}$ are continuous and require, consequently, that the differential equation (6.47) is satisfied pointwise at each point of G.

6.3.2. If the coefficients $a(z)$ and $b(z)$ are equal to zero at each point z of G, then the equation (6.47) passes into

$$\frac{\partial w}{\partial \bar{z}} = 0,$$

i.e., one obtains the differential equation (6.13) for holomorphic functions. Thus in this case generalized analytic functions turn out to be holomorphic. In view of Weyl's lemma (see 6.1.6.) the latter statement is true for any (integrable) solution in Sobolev's sense to (6.47) provided $a(z)$ and $b(z)$ equal to zero everywhere. In general, i.e., in the case that $a(z)$ and $b(z)$ are not equal to zero everywhere, it is possible that there exist solutions in Sobolev's sense which are not classical solutions (such example is given in the following section 6.3.3.).

6.3.3. Regard the function $w = w(z)$ defined by

$$w(z) = \exp\left(-2z \ln \ln \frac{1}{r}\right), \tag{6.48}$$

where $z = x + iy = r\exp(i\theta)$. Since $r \cdot \ln \ln \frac{1}{r}$ tends to zero as r tends to zero, the function $w = w(z)$ is also continuous at the point $z = 0$ provided we define $w(0) = 1$. Using the ordinary chain rule, one gets immediately

$$\frac{\partial w}{\partial x}(z) = \exp\left(-2z \ln \ln \frac{1}{r}\right)\left(-2 \ln \ln \frac{1}{r} + \frac{2zx}{r^2 \ln \frac{1}{r}}\right)$$

and

$$\frac{\partial w}{\partial y}(z) = \exp\left(-2z \ln \ln \frac{1}{r}\right)\left(-2 \ln \ln \frac{1}{r} + \frac{2zy}{r^2 \ln \frac{1}{r}}\right).$$

In view of the definition (6.4) we obtain

$$\frac{\partial w}{\partial \bar z}(z) = \exp(-2z \ln \ln \frac{1}{r}) \frac{z^2}{r^2 \ln \frac{1}{r}} . \tag{6.49}$$

Hence the derivative $\partial w/\partial \bar z$ possesses the limit 0 at the point $z = 0$, although the partial derivatives $\partial w/\partial x$ and $\partial w/\partial y$ are unbounded at the same point. From the last formula (6.49) it follows, further, that the function (6.48) is a solution to the differential equation (6.47) in the case

$$a(z) = \frac{\exp(2i\theta)}{\ln \frac{1}{r}}, \quad b(z) = 0.$$

Since these coefficients $a(z)$, $b(z)$ are continuous everywhere (provided we define $a(0) = 0$) the function (6.48) turns out to be a non-classical solution to a differential equation with continuous coefficients.

6.3.4. The book [77] of I. N. Vekua contains a general theory of the differential equation (6.47) under the assumption that the coefficients $a(z)$, $b(z)$ belong only to the space $L_p(G)$, $p > 2$. The solutions looked for are continuous (with the exception of isolated singularities). The derivative $\partial w/\partial \bar z$ itself belongs to $L_p(G)$, too, and the differential equation is satisfied almost everywhere.

6.3.5. It may be added that each uniformly elliptic linear system for two desired real-valued functions in the plane can be reduced to the system (6.47), which may be interpreted as canonical form of those systems (cf. also section 8.1.1.). Nonlinear systems cannot be written in this form (6.47), in general, but they can be reduced to equations of the type

$$\frac{\partial w}{\partial \bar z} = F(z, w, \frac{\partial w}{\partial z}) . \tag{6.50}$$

The same is true in the case that w is a vector satisfying a linear system of differential equations. The equation (6.50) behaves like an elliptic one if the right-hand side satisfies a Lipschitz condition with respect to $\partial w/\partial z$ with a sufficiently small Lipschitz constant, and then many problems such as boundary value problems are solvable as in the linear case (cf. [64]). This is the reason why sometimes the solutions to (6.50) are also called generalized analytic functions. The form (6.50) is obtained by solving a given system for the derivative $\partial w/\partial \bar z$. Partial complex differential equations not solved for $\partial w/\partial \bar z$ are investigated by W. Rüprich [55, 56].

6.3.6. Let $w = w(z)$ be a given continuous solution to (6.47) in G, where $a(z)$ and $b(z)$ are defined and continuous in G. Let K, further, be any compact subset of G. Now choose a domain G' containing K and having the property that $\overline{G'}$ is a compact subset of G. Then define Φ by

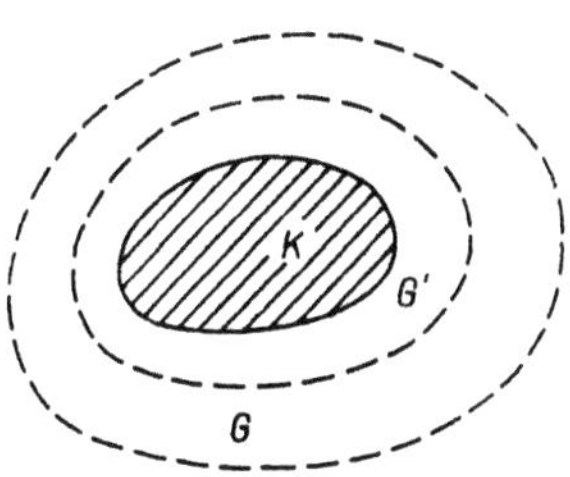

$$\Phi = w - T_{G'}(aw + b\overline{w}). \tag{6.51}$$

Taking into account formula (6.17), one obtains

$$\frac{\partial\Phi}{\partial\overline{z}} = \frac{\partial w}{\partial\overline{z}} - (aw + b\overline{w}) = 0 \quad \text{in } G'$$

since w is a solution to (6.47). Thus the function Φ turns out to be holomorphic in G'. In view of 6.2.9. the function Φ is Hölder-continuous in K. On the other hand the function $aw + b\overline{w}$ is continuous in $\overline{G'}$ and belongs, therefore, to each $L_p(G')$, $p > 2$, especially. According to 6.2.12. the operator $T_{G'}$ maps $L_p(G')$ into $C^{\beta}(\overline{G'})$, $\beta = \frac{p-2}{p}$. Hence the function $T_{G'}(aw + b\overline{w})$ is proved to be Hölder-continuous with the exponent β in K. If we choose $p = \frac{2}{1-\alpha}$, we get an arbitrarily prescribed Hölder-exponent α, $0 < \alpha < 1$. Since (6.51) may be rewritten as

$$w = T_{G'}(aw + b\overline{w}) + \Phi \tag{6.52}$$

we have proved the following result:

Assume the coefficients of (6.47) are continuous. Then every continuous solution $w = w(z)$ is necessarily Hölder-continuous with the arbitrarily chosen exponent α, $0 < \alpha < 1$, in each compact subset K of G.

If the coefficients $a(z)$, $b(z)$ belong to $L_p(G)$, $p > 2$, then every continuous solution is Hölder-continuous with the exponent $\beta = \frac{p-2}{p}$.

<u>6.3.7.</u> Let G be a domain in the z-plane and K a compact subset of G. Suppose that the coefficients of (6.47) are Hölder-continuous with the exponent α, $0 < \alpha < 1$. Then the following theorem holds:

<u>Theorem.</u> Any continuous solution $w = w(z)$ to (6.47) in G belongs to $C^{1+\alpha}(K)$.

<u>Proof.</u> Choose the domain G' in a similar way as done in 6.3.6. Applying 6.3.6. to $\overline{G'}$, we see that w is Hölder-continuous with the exponent α in $\overline{G'}$ because $\overline{G'}$ itself is a compact subset of G (i.e., we replace the set K in 6.3.6. by $\overline{G'}$. The domain G' in 6.3.7. is to be replaced, consequently, by a larger one). Then define Φ by (6.51). Hence w is representable by (6.52). Differentiating this representation with respect to both variables z and $\overline{z}$, one obtains

$$\begin{aligned} \frac{\partial w}{\partial z} &= \Pi_{G'}(aw + b\overline{w}) + \Phi', \\ \frac{\partial w}{\partial\overline{z}} &= aw + b\overline{w}. \end{aligned} \tag{6.53}$$

Taking into account that the product and the sum of Hölder-continuous functions are Hölder-continuous again, we get that $aw + b\overline{w}$ is Hölder-continuous in $\overline{G'}$. By virtue of 6.2.14. the term $\Pi_{G'}(aw + b\overline{w})$ is Hölder-continuous in $\overline{G'}$, too. Since also $T_{G'}(aw + b\overline{w})$ is Hölder-continuous in $\overline{G'}$ in view of 6.2.7., the function Φ defined by (6.51) is Hölder-continuous in $\overline{G'}$, too. Moreover, according to 6.2.9. the derivative Φ' turns out to be Hölder-continuous in K. Therefore the formulae (6.53) show that both derivatives $\partial w/\partial z$ and $\partial w/\partial \overline{z}$ are Hölder-continuous with the exponent α in K. This completes the proof of the theorem.

The proof is based on the fact that $T_{G'}$ and $\Pi_{G'}$ are linear and bounded operators mapping $C^{\alpha}(\overline{G'})$ into itself. The same is true in the case of the space of all functions which are n times Hölder-continuously differentiable in $\overline{G'}$. Thus the formulae (6.52) and (6.53) allow one also to prove that w is necessarily $(n + 1)$ times Hölder-continuously differentiable in K provided the coefficients a(z), b(z) are n times Hölder-continuously differentiable in G.

<u>6.3.8.</u> Let G be a bounded domain in the z-plane. Suppose that the coefficients a(z), b(z) of (6.47) are Hölder-continuous in $\overline{G}$ with the exponent α, $0 < \alpha < 1$. Denote the space of all the solutions to (6.47) which are Hölder-continuous with a fixed exponent α in $\overline{G}$ by W, i.e., W consists of those elements of $C^{\alpha}(\overline{G})$ which satisfy the differential equation (6.47) in G.

<u>Theorem.</u> W is a Banach space.

Since $C^{\alpha}(\overline{G})$ is a Banach space (cf. 6.2.6.) it remains to prove that the limit of a fundamental sequence $\{w_n\}_{n=1,2,\ldots}$ in W belongs to W, too. Analogously to (6.51) define Φ_n by

$$\Phi_n = w_n - T_G(aw_n + b\overline{w}_n). \tag{6.54}$$

Since every w_n is a solution to (6.47) it follows

$$\frac{\partial \Phi_n}{\partial \overline{z}} = \frac{\partial w_n}{\partial \overline{z}} - (aw_n + b\overline{w}_n) = 0$$

and, consequently, each Φ_n is holomorphic in G (cf. also 6.3.6.).

Further take into account that convergence in $C^{\alpha}(\overline{G})$ implies uniform convergence. Denote the limit function of the w_n by w_*. In view of (6.54) we have

$$\Phi_n - \Phi_m = (w_n - w_m) - T_G\left[a(w_n - w_m) + b\overline{(w_n - w_m)}\right]$$

and, therefore,

$$\|\Phi_n - \Phi_m\|_{C^{\alpha}(\overline{G})}$$
$$\leq \|w_n - w_m\|_{C^{\alpha}(\overline{G})} + \|T_G\| \cdot \|a(w_n - w_m) + b\overline{(w_n - w_m)}\|_{C^{\alpha}(\overline{G})}$$

because T_G is a bounded operator. Especially, this estimate shows that the Φ_n converge uniformly. Thus the limit function Φ_* is holomorphic in G in view of Weierstrass' convergence theorem. Carrying out the limiting process $n \to \infty$ in (6.54) under the sign of the integral, we get the relation

$$\Phi_* = w_* - T_G(aw_* + b\bar{w}_*).$$

Differentiating the last equation with respect to $\bar{z}$ and taking into consideration the holomorphy of Φ_*, we get

$$0 = \frac{\partial w_*}{\partial \bar{z}} - (aw_* + b\bar{w}_*).$$

Consequently, the limit function w_* is a solution to (6.47) in G.

6.3.9. Let G be a bounded domain in the z-plane and G' any subdomain having the positive distance δ from the boundary of G. Suppose that the coefficients a(z), b(z) of (6.47) are Hölder-continuous with the exponent α in $\bar{G}$. Take any solution $w = w(z)$ belonging to the Banach space W defined in the preceding section 6.3.8. Applying the theorem of 6.3.7., we see that $\partial w/\partial z$ is Hölder-continuous in $\overline{G'}$, especially. The following theorem allows us to estimate the norm of the derivative $\partial w/\partial z$ in $\overline{G'}$ by that of w.

Theorem. There exists a constant C not depending on G' such that for each solution w to (6.47) the estimate

$$\left\|\frac{\partial w}{\partial z}\right\|_{C^\alpha(\overline{G'})} \leq \frac{C}{\delta}\|w\|_{C^\alpha(\bar{G})}$$

holds.

Proof. Let w be an arbitrary solution belonging to W. Then define the holomorphic function Φ, depending on the choice of w, by

$$\Phi = w - T_G(aw + b\bar{w}) \tag{6.55}$$

(cf. 6.3.6., especially (6.51)). According to 6.2.7. the function Φ belongs to $C^\alpha(\bar{G})$. In order to estimate $\|\Phi\|_{C^\alpha(\bar{G})}$ by the norm $\|w\|_{C^\alpha(\bar{G})}$, we take into consideration, first, that the product w_1w_2 of two functions belonging to $C^\alpha(\bar{G})$ is an element of $C^\alpha(\bar{G})$, again. Using the norm in $C^\alpha(\bar{G})$ defined in 6.2.6., it is easy to prove that

$$\|w_1w_2\|_{C^\alpha(\bar{G})} \leq 2\|w_1\|_{C^\alpha(\bar{G})}\|w_2\|_{C^\alpha(\bar{G})}.$$

Therefore one gets

$$\|aw + b\bar{w}\|_{C^\alpha(\bar{G})} \leq C_1\|w\|_{C^\alpha(\bar{G})}, \tag{6.56}$$

where for short we use the abbreviation C_1 for the constant

$$2(\|a\|_{C^\alpha(G)} + \|b\|_{C^\alpha(\bar{G})}).$$

By virtue of the definition (6.55) of Φ we obtain, consequently, the desired estimate

$$\|\Phi\|_{C^\alpha(\overline{G})} \leq (1 + \|T_G\| C_1)\|w\|_{C^\alpha(\overline{G})}. \tag{6.57}$$

Applying 6.2.11. we get, further, the estimate

$$\|\Phi'\|_{C^\alpha(\overline{G'})} \leq \frac{C_2}{\delta}\|\Phi\|_{C^\alpha(\overline{G})},$$

where

$$C_2 = 3\cdot 2^\alpha$$

(cf. (6.26)). Thus the inequality (6.57) yields

$$\|\Phi'\|_{C^\alpha(\overline{G'})} \leq \frac{1}{\delta}(1 + \|T_G\| C_1)C_2\|w\|_{C^\alpha(\overline{G})}. \tag{6.58}$$

Differentiating (6.55) with respect to z, one obtains

$$\frac{\partial w}{\partial z} = \Phi' + \Pi_G(aw + b\overline{w}).$$

Since Π_G is a bounded operator mapping $C^\alpha(\overline{G})$ into itself (see 6.2.14.), the last equation leads to the estimate

$$\left\|\frac{\partial w}{\partial z}\right\|_{C^\alpha(\overline{G'})} \leq \|\Phi'\|_{C^\alpha(\overline{G'})} + \|\Pi_G\|\cdot\|aw + b\overline{w}\|_{C^\alpha(\overline{G})},$$

where the $C^\alpha(\overline{G'})$-norm of the second term on the right-hand side is replaced by the $C^\alpha(\overline{G})$-norm, which is not smaller than the first one. In view of (6.56) and (6.58) the last estimate gives

$$\left\|\frac{\partial w}{\partial z}\right\|_{C^\alpha(\overline{G'})} \leq \left(\frac{1}{\delta}(1 + \|T_G\| C_1)C_2 + \|\Pi_G\| C_1\right)\|w\|_{C^\alpha(\overline{G})}.$$

Denote the diameter of G by diam G. Then the distance of any subdomain G' of G from the boundary of G is not larger than diam G. Hence one gets the final estimate

$$\left\|\frac{\partial w}{\partial z}\right\|_{C^\alpha(\overline{G'})} \leq \frac{C}{\delta}\|w\|_{C^\alpha(\overline{G})},$$

where

$$C = (1 + \|T_G\| C_1)C_2 + \|\Pi_G\| C_1 \text{ diam } G.$$

6.4. Associated differential operators

The ordinary complex derivative $\frac{dh}{dz}$ of a holomorphic function h is holomorphic, again, i.e., the complex differentiation $\frac{d}{dz}$ is an operator mapping the space H of holomorphic functions (cf. section 2.1.) into itself. In section 6.3.8. we introduced the space W consisting of all solutions to the differential equation (6.47). On the other hand the differential equation (6.13) for holomorphic functions is a special case of (6.47), cf. section 6.3.2. Therefore the space H turns out to be identical with the space W in the case that a(z) and b(z) are equal to zero everywhere.

The question whether there exist first order differential operators mapping the space W into itself leads to the concept of associated differential operators which we define now.

6.4.1. Regard the differential equation

$$lw = 0,$$

where l is a given differential operator. Then a second differential operator L is called associated to l if L transforms the set of all solutions to the differential equation $lw = 0$ into itself, i.e., if

$$lw = 0 \quad \text{implies} \quad l(Lw) = 0.$$

The space H of all holomorphic functions defined in a fixed domain may be characterized by the differential equation (6.13). Thus the operator l is

$$l = \frac{\partial}{\partial \bar{z}}$$

in this case. In view of 6.1.5. the complex derivative $\frac{dw}{dz}$ of a holomorphic function $w = w(z)$ may be written in the form $\frac{\partial w}{\partial z}$. Since $\frac{\partial w}{\partial z}$ is holomorphic again we have

$$\frac{\partial}{\partial \bar{z}}\left(\frac{\partial w}{\partial z}\right) = 0.$$

Thus $L = \frac{\partial}{\partial z}$ is associated to $l = \frac{\partial}{\partial \bar{z}}$.

6.4.2. Now assume that $lw = 0$ is identical with the differential equation (6.47), i.e., we take

$$lw = \frac{\partial w}{\partial \bar{z}} - a(z)w - b(z)\bar{w}. \tag{6.59}$$

In the special case that a(z) and b(z) are equal to zero everywhere we know already that $L = \frac{\partial}{\partial z}$ is associated to l. Generalizing this example, we look for operators L associated to l which have the form

$$Lw = C(z)\frac{\partial w}{\partial z} + A(z)w + B(z)\bar{w}. \tag{6.60}$$

For a given operator l of type (6.59) we try to find sufficient conditions on A(z), B(z), and C(z) under which L turns out to be associated to l. For this end we regard the expression

$$\begin{aligned} l(Lw) &= \frac{\partial}{\partial \bar{z}}(Lw) - a(Lw) - b\overline{(Lw)} \\ &= \frac{\partial}{\partial \bar{z}}(C\frac{\partial w}{\partial z} + Aw + B\bar{w}) \\ &\quad - a(C\frac{\partial w}{\partial z} + Aw + B\bar{w}) - b(\bar{C}\overline{\left(\frac{\partial w}{\partial z}\right)} + \bar{A}\bar{w} + \bar{B}w). \end{aligned}$$

Provided the function $w = w(z)$ is twice continuously differentiable, it follows easily from the definitions (6.3) and (6.4) that

$$\frac{\partial}{\partial \bar{z}}(\frac{\partial w}{\partial z}) = \frac{1}{4}(\frac{\partial}{\partial x} + i\frac{\partial}{\partial y})(\frac{\partial w}{\partial x} - i\frac{\partial w}{\partial y}) = \frac{1}{4}(\frac{\partial^2 w}{\partial x^2} + \frac{\partial^2 w}{\partial y^2}),$$

$$\frac{\partial}{\partial z}\left(\frac{\partial w}{\partial \bar z}\right) = \frac{1}{4}\left(\frac{\partial}{\partial x} - i\frac{\partial}{\partial y}\right)\left(\frac{\partial w}{\partial x} + i\frac{\partial w}{\partial y}\right) = \frac{1}{4}\left(\frac{\partial^2 w}{\partial x^2} + \frac{\partial^2 w}{\partial y^2}\right).$$

Comparing both formulae, we see that

$$\frac{\partial}{\partial \bar z}\left(\frac{\partial w}{\partial z}\right) = \frac{1}{4}\Delta w = \frac{\partial}{\partial z}\left(\frac{\partial w}{\partial \bar z}\right). \tag{6.61}$$

If $w = w(z)$ is a solution to $lw = 0$, i.e., to (6.47), we have

$$\frac{\partial}{\partial z}\left(\frac{\partial w}{\partial \bar z}\right) = \frac{\partial}{\partial z}(aw + b\bar w).$$

Using the differentiation rules 6.1.2., one obtains

$$\frac{\partial}{\partial z}\left(\frac{\partial w}{\partial \bar z}\right) = a\frac{\partial w}{\partial z} + \frac{\partial a}{\partial z}w + b\left(\overline{\frac{\partial w}{\partial \bar z}}\right) + \frac{\partial b}{\partial z}\bar w.$$

Once more taking into account the differential equation $lw = 0$, we get further

$$\frac{\partial}{\partial z}\left(\frac{\partial w}{\partial \bar z}\right) = a\frac{\partial w}{\partial z} + \left(\frac{\partial a}{\partial z} + b\bar b\right)w + \left(\bar a b + \frac{\partial b}{\partial z}\right)\bar w. \tag{6.62}$$

In view of (6.61) the second order derivative $\frac{\partial}{\partial \bar z}\left(\frac{\partial w}{\partial z}\right)$ is also equal to the right-hand side of (6.62). Once more applying the differentiation rules 6.1.2. and replacing $\frac{\partial w}{\partial \bar z}$ by $aw + b\bar w$, one obtains, therefore, that $l(Lw)$ is a linear combination of

$$\frac{\partial w}{\partial z},\quad \overline{\left(\frac{\partial w}{\partial z}\right)},\quad w,\quad \text{and}\quad \bar w,$$

where the coefficients are

$$\frac{\partial C}{\partial \bar z},\quad B - b\bar C,\quad \frac{\partial a}{\partial z}C + \frac{\partial A}{\partial \bar z} + b\bar b c - b\bar B,\quad \text{and}\quad \frac{\partial b}{\partial z}C + \frac{\partial B}{\partial \bar z} + \bar a b C - aB \text{ resp.}$$

Equating these four expressions to zero, we get sufficient conditions on the coefficients a, b, A, B, and C under which the equation $lw = 0$ implies $l(Lw) = 0$. The first equation means that C is holomorphic. The second equation is

$$B = b\bar C. \tag{6.63}$$

Substituting this relation, the third and the fourth equation pass into

$$\frac{\partial a}{\partial z}C + \frac{\partial A}{\partial \bar z} = 0 \tag{6.64}$$

$$\frac{\partial b}{\partial z}C + \frac{\partial b}{\partial \bar z}\bar C + b\overline{\left(\frac{dC}{dz}\right)} + b(\bar a C - a\bar C) + b(A - \bar A) = 0 \tag{6.65}$$

since C is holomorphic and, consequently,

$$\frac{\partial \bar C}{\partial \bar z} = \overline{\left(\frac{\partial C}{\partial z}\right)} = \overline{\left(\frac{dC}{dz}\right)}$$

(see section 6.1.2.). Summarizing these calculations, the following theorem has been proved:

Theorem. Suppose that $C = C(z)$ is holomorphic and further that the coefficients $a(z)$, $b(z)$, $A(z)$, $B(z)$, and $C(z)$ of l and L satisfy the relations (6.63), (6.64), and (6.65). Then the differential operator L is associated to l.

6.4.3. Now we assume that the differential operator l of type (6.59) is given. Then we look for an operator L of type (6.60) associated to l. In order to construct such operator, we try to find coefficients $A(z)$, $B(z)$, and $C(z)$ satisfying the three equations (6.63), (6.64), and (6.65). According to the theorem in the preceding section 6.4.2., we have to choose $C = C(z)$ as holomorphic function. Take

$$C(z) = 1. \tag{6.66}$$

Then relation (6.63) states that

$$B = b. \tag{6.67}$$

Substituting (6.66) and (6.67) into (6.64) and (6.65), we see that the remaining coefficient $A(z)$ must satisfy the two equations

$$\frac{\partial a}{\partial z} + \frac{\partial A}{\partial \bar{z}} = 0, \tag{6.68}$$

$$\frac{\partial b}{\partial z} + \frac{\partial b}{\partial \bar{z}} + b(\bar{a} - a) + b(A - \bar{A}) = 0. \tag{6.69}$$

at the same time. Thus $A(z)$ has to satisfy the inhomogeneous Cauchy-Riemann equation (6.68) with a linear algebraic side condition (6.69). The system (6.68), (6.69) turns out to be overdetermined. Hence this system is solvable only if the given coefficients $a(z)$, $b(z)$ satisfy a compatibility condition. In order to deduce this condition, we split up the equations (6.68) and (6.69) into their real and imaginary parts. Denote the real part of A by A_1 and the imaginary part by A_2 and so on, i.e., introduce the denotations

$$\begin{aligned} A &= A_1 + iA_2, \quad B = B_1 + iB_2, \\ a &= a_1 + ia_2, \quad b = b_1 + ib_2. \end{aligned} \tag{6.70}$$

Taking into account the relation

$$\frac{\partial}{\partial z} + \frac{\partial}{\partial \bar{z}} = \frac{\partial}{\partial x}$$

which follows immediately from the definitions (6.3) and (6.4), the complex equation (6.69) yields the two equations

$$\mathrm{Re}\left[\frac{1}{b}\,\frac{\partial b}{\partial x}\right] = 0 \tag{6.71}$$

and

$$A_2 = a_2 - \frac{1}{2}\,\mathrm{Im}\left[\frac{1}{b}\,\frac{\partial b}{\partial x}\right] \tag{6.72}$$

provided the coefficient b is different from zero everywhere. Now suppose that the given coefficient b can be written in the form

$$b = \exp b'.$$

This assumption is satisfied, for instance, if $b(z)$ ($\neq 0$ everywhere) is defined in a simply connected domain since in this case each branch of $b' = \log b$ is uniquely defined. Similarly as in (6.70) we denote the real and imaginary part of b' by b_1' and b_2' resp., i.e., we set $b' = b_1' + ib_2'$. Then the equation (6.71) passes into

$$\frac{\partial b_1'}{\partial x} = 0 \tag{6.73}$$

while (6.72) may be rewritten as

$$A_2 = a_2 - \frac{1}{2}\frac{\partial b_2'}{\partial x}. \tag{6.74}$$

Once more taking into consideration the definitions (6.3) and (6.4), by splitting up (6.68) into real and imaginary part it follows

$$\frac{\partial A_1}{\partial x} = \frac{\partial A_2}{\partial y} - \frac{\partial a_1}{\partial x} - \frac{\partial a_2}{\partial y},$$

$$\frac{\partial A_1}{\partial y} = -\frac{\partial A_2}{\partial x} - \frac{\partial a_2}{\partial x} + \frac{\partial a_1}{\partial y}.$$

Substituting (6.74) into this system, we obtain the first order system

$$\begin{aligned} \frac{\partial A_1}{\partial x} &= -\frac{1}{2}\frac{\partial^2 b_2}{\partial x \partial y} - \frac{\partial a_1}{\partial x}. \\ \frac{\partial A_1}{\partial y} &= +\frac{1}{2}\frac{\partial^2 b_2'}{\partial x^2} - 2\frac{\partial a_2}{\partial x} + \frac{\partial a_1}{\partial y}. \end{aligned} \tag{6.75}$$

We suppose that all derivatives appearing in our calculations exist and are continuous. In order to ensure that a system of type (6.75) possesses at least one solution, we must suppose that the right-hand sides satisfy the well-known integrability condition. In order to rewrite this condition, we equate the right-hand side of the first equation differentiated to y with that of the second equation differentiated to x. In this way one obtains

$$\frac{\partial}{\partial x}\left[4\left(\frac{\partial a_2}{\partial x} - \frac{\partial a_1}{\partial y}\right) - \Delta b_2'\right] = 0.$$

By using the formula (6.61) and the representation of $\partial a/\partial z$ by real derivatives of a_1 and a_2, the real differentiations in the brackets may be replaced by complex ones. Hence the last condition leads to

$$\frac{\partial}{\partial x}\left[2\ \mathrm{Im}\frac{\partial a}{\partial z} - \frac{\partial^2 b'}{\partial z \partial \bar{z}}\right] = 0, \tag{6.76}$$

where we took into account that in view of (6.73) the imaginary part b_2' can be replaced by the complex-valued function b'. If the condition (6.76) is satisfied in a given simply connected domain G, then there exists a function $A_1 = A_1(x,y)$ satisfying the system (6.76). It may be added that the solution A_1 is uniquely determined up to a real constant. We sum up our result in the following theorem:

Theorem. Suppose that the coefficients a(z) and b(z) of the differential operator l,

$$lw = \frac{\partial w}{\partial \bar{z}} - a(z)w - b(z)\bar{w},$$

are defined in the domain G, where $b(z) \neq 0$ everywhere. Suppose, moreover, that both functions

$$\ln|b|$$

and

$$2\left[\mathrm{Im}\,\frac{\partial a}{\partial z} - \frac{\partial^2 \log b}{\partial z\,\partial \bar{z}}\right]$$

do not depend on x. If G is simply connected, then there exist coefficients A(z) and B(z) such that the operator L,

$$Lw = \frac{\partial w}{\partial z} + A(z)w + B(z)\bar{w},$$

is associated to l.

Finally we would like to remark that we may look for such operators L of type (6.60) for which $C = C(z)$ is any given holomorphic function in G. Instead of satisfying the conditions (6.73) and (6.76), the coefficients a(z) and b(z) must satisfy similar conditions depending on the choice of C(z).

6.4.4. If solutions in Sobolev's sense are included into the considerations, the differential equation in question must not necessarily be satisfied pointwise. Solutions belonging to L_p, for instance, have to satisfy the differential equation almost everywhere. Piecewise continuously differentiable solutions possessing jumps along a finite number of curves must not satisfy the differential equation on the curves themselves. If we look for associated differential operators of type (6.59) and (6.60) with piecewise constant coefficients, the differential equations (6.68) and (6.69) need not be satisfied on the curves along which jumps are permitted. Outside the curves the differential equation (6.68) is alsways satisfied if the coefficients are constant, whereas the differential equation (6.69) passes into

$$b(\bar{a} - a) + b(A - \bar{A}) = 0,$$

i.e.,

$$\mathrm{Im}\,A = \mathrm{Im}\,a$$

provided $b \neq 0$. Choosing $C(z) = 1$ everywhere and taking into considerations the relation (6.67), the following result has been proved:

Let l and L be differential operators with piecewise constant coefficients,

$$lw = \frac{\partial w}{\partial \bar{z}} - aw - b\bar{w}, \qquad Lw = \frac{\partial w}{\partial z} + Aw + B\bar{w}.$$

The differential operator L is associated to l if

$$\operatorname{Im} A = \operatorname{Im} a \quad \text{and} \quad B = b.$$

Example. These conditions are satisfied if $A = 0$, $B = 1$, $a = 0$, $b = 1$. Thus the differential operator L with $Lw = \frac{\partial w}{\partial z} + \bar{w}$ is associated to l, $lw = \frac{\partial w}{\partial \bar{z}} - \bar{w}$.

Let us additionally remark that in the case $b = 0$ the constant A may be chosen arbitrarily.

6.4.5. Now return to the differential operators (6.59) and (6.60) in the case that the coefficients depend on z. Calculations analogous to those of section 6.4.2. yield the following statement:

If the relations

$$B = bC,$$

$$\frac{\partial a}{\partial z} + \frac{\partial}{\partial \bar{z}}\left(\frac{A}{C}\right) = 0,$$

$$\left[\frac{\partial b}{\partial z} + \frac{\partial}{\partial \bar{z}}\left(\frac{B}{C}\right)\right]C + (\bar{a} - a)B + bA - \frac{B\bar{A}}{\bar{C}} = 0$$

are satisfied ($C(z) \neq 0$ everywhere), then the differential operator l is associated to L. In the case $C(z) = 1$ (everywhere) these conditions pass into (6.67), (6.68), and (6.69). Hence those relations (6.67), (6.68), (6.69) do not only guarantee that L is associated to l, but they ensure also that l is associated to L.

6.5. Differentiability properties of associated differential operators

Let us once more start from the fact that the complex differentiation $\frac{\partial}{\partial z}$ transforms each holomorphic function $w = w(z)$ into a further holomorphic function $\frac{dw}{dz}$. Repeating this consideration, it follows that there exist all derivatives $\frac{d^k w}{dz^k}$, $k = 1, 2, \ldots$, too. On the other hand, the operator L transforms the set W consisting of the set of all the solutions to $lw = 0$ into itself provided L is associated to l. Thus all the iterated images $L^k w$, $k = 1, 2, \ldots$, must exist, too. In the present section the question is investigated under which conditions on $a(z)$, $b(z)$ all the images $L^k w$ turn out to be Hölder-continuous.

Notice, first, that in view of 6.3.7. any continuous solution to the differential equation (6.47) is Hölder-continuously differentiable in any subdomain G' of G provided the coefficients $a(z)$, $b(z)$ themselves are Hölder-continuous in G and G' is a compact subset of G (we apply 6.3.7. with $K = \overline{G'}$). If the coefficients a(z), b(z) are Hölder-continuously differentiable, then w is proved to be twice Hölder-continuously differentiable in G' (cf. the corresponding statement at the end of section 6.3.7.).

Now assume that the coefficients A(z), B(z), and C(z) of L (see (6.60)) are Hölder-continuously differentiable and connected with a(z) and b(z) by the equations (6.67), (6.68), and (6.69). According to the theorem in 6.4.2., the coefficient C(z), moreover, is assumed to be holomorphic. From the definition (6.60) of Lw it follows immediately that Lw is Hölder-continuously differentiable. Substituting Lw into the differential equation (6.47), the calculations in 6.4.2. show that Lw is a Hölder-continuously differentiable solution to (6.47). Repeating the above considerations that are based on the section 6.3.7., we see that Lw is twice Hölder-continuously differentiable and, consequently, $L^2w = L(Lw)$ is at least once Hölder-continuously differentiable. By induction one obtains the following theorem:

<u>Theorem.</u> Assume that the coefficients a(z), b(z), A(z), B(z) are Hölder-continuously differentiable and C(z) is holomorphic. Then for every k, k = 1, 2, ..., the iterated images L^kw exist and are Hölder-continuously differentiable for any solution w to the equation lw = 0.

Notice, further, that again in view of 6.3.7. every solution to (6.47) is (n + 1) times Hölder-continuously differentiable provided the coefficients a(z), b(z) are n times Hölder-continuously differentiable. Therefore, all L^kw, k = 1, 2, ..., are (n + 1) times Hölder-continuously differentiable provided the coefficients a(z) and b(z) are n times Hölder-continuously differentiable.

Although L^K is a differential operator of the order k, the Hölder-continuity of L^Kw does not imply, of course, that all partial derivatives of w of order k do exist and are Hölder-continuous.

Finally assume that L is any differential operator associated to l, not necessarily connected with l by the relations (6.67), (6.68), and (6.69). We assume only that the coefficients of L as well as those of l are Hölder-continuous. Since in view of 6.3.7. every solution to (6.47) is Hölder-continuously differentiable we get by induction that L^Kw is Hölder-continuously differentiable for each k, k = 1, 2, ..., provided w is any (continuous) solution to the differential equation lw = 0.

7. INITIAL VALUE PROBLEMS WITH GENERALIZED ANALYTIC INITIAL FUNCTIONS

7.1. Statement of the problem

Using the method of scales of Banach spaces, in chapter 4 we have solved initial value problems of the type (0.3), (0.2) in the special case that the right-hand side of the differential equation as well as the initial function are holomorphic. In the present chapter we investigate initial value problems in case that the right-hand side does not transform the space of all holomorphic functions into itself. Accordingly, the initial function need not be holomorphic.

We shall see that there is a close connection beween possible right-hand sides and possible initial functions: The initial value problem

$$\frac{\partial w}{\partial t} = \tilde{L}w, \tag{7.1}$$

$$w(0,z) = w_0(z) \tag{7.2}$$

turns out to be solvable if the initial function w_0 is a solution to the differential equation $lw = 0$ and the right-hand side is associated to l.

In section 7.4. solutions to the initial value problem (7.1), (7.2) will be constructed in the case that the desired function w is complex-valued and depends on both the real variable t and the complex variable $z = x + iy$. The operator l will be defined by (6.59), i.e., the differential equation $lw = 0$ for the permissible initial functions is identical with (6.47). The permissible right-hand sides $\tilde{L}$ must transform, consequently, the space W of generalized analytic functions defined by (6.47) (cf. 6.3.8.) into itself.

In section 6.4.2. sufficient conditions are deduced under which operators L of type (6.60) transform the above-mentioned space W into itself. Therefore we shall investigate initial value problems of type (7.1), (7.2) in the case that the right-hand side of (7.1) is given by the right-hand side of (6.60), especially. We permit, as done in chapter 4, that the coefficients of the right-hand sides depend explicitly on the variable t, i.e., generalizing right-hand sides of type (6.60), in the section 7.4. we shall solve initial value problems (7.1), (7.2) with right-hand sides $\tilde{L}w$ of the form

$$\tilde{L}w = C(t,z)\frac{\partial w}{\partial z} + A(t,z)w + B(t,z)\overline{w} + D(t,z).$$

We would like to emphasize that the coefficients $a(z)$, $b(z)$ of l, however, must not depend on t.

In section 6.4.3. we determined operators L associated to a given operator l of type (6.59). From the point of view of the theory of initial value problems of type (7.1), (7.2) we have, conversely, to look for operators l to which a given operator is associated. This inverse problem will be solved in section 7.3. These considerations will be prepared by the next section dealing with overdetermined first order systems whose right-hand sides depend on the desired function.

7.2. A lemma on an overdetermined first order system

Let G be a convex domain in the x,y-plane. Without any loss of generality we may assume that the origin (0,0) belongs to G. Then we look for a real-valued function $u = u(x,y)$ satisfying in G the real first order system

$$\frac{\partial u}{\partial x} = f_1(x,y,u), \qquad \frac{\partial u}{\partial y} = f_2(x,y,u) \tag{7.3}$$

and taking the arbitrarily prescribed value u_0 at the origin, i.e., the desired solution has to satisfy the initial condition

$$u(0,0) = u_0, \tag{7.4}$$

too. The right-hand sides $f_1(x,y,u)$, $f_2(x,y,u)$ of (7.3) are continuously differentiable functions defined for every (x, y) belonging to G and for each real u.

If the right-hand sides f_1, f_2 do not depend on u, then the existence of at least one twice continuously differentiable solution $u = u(x,y)$ implies that

$$\frac{\partial^2 u}{\partial x \partial y} = \frac{\partial f_1}{\partial y}, \quad \frac{\partial^2 u}{\partial y \partial x} = \frac{\partial f_2}{\partial x}$$

(this follows immediately from (7.3)). Consequently, the right-hand sides f_1, f_2 must necessarily satisfy the relation

$$\frac{\partial f_1}{\partial y} = \frac{\partial f_2}{\partial x}.$$

Now return to the general case, in which the right-hand sides do depend on the sought solution. The solution $u = u(x,y)$ to the initial value problem (7.3), (7.4) is easily obtainable if there is any. In order to construct this solution, we denote $u(x,0)$ by $\phi(x)$. Then ϕ has to satisfy the ordinary differential equation

$$\frac{d\phi}{dx} = f_1(x,0,\phi) \tag{7.5}$$

which is nothing else than the first equation (7.3) specialized for points of the real axis. Thus $\phi(x)$ is the uniquely determined solution to (7.5) satisfying the initial condition $\phi(0) = u_0$. Interpreting the

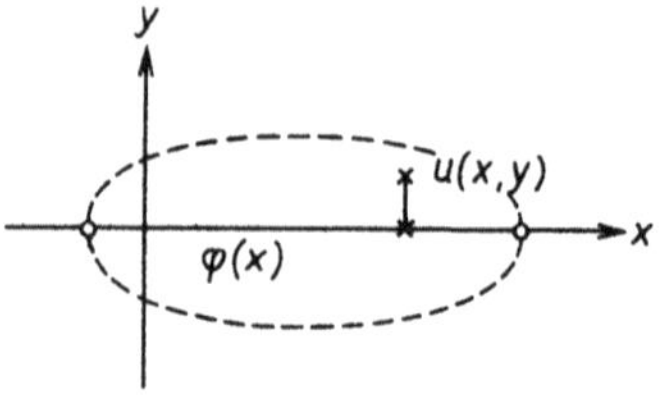

second equation (7.3) as ordinary differential equation with respect to y (for fixed x), one obtains u(x,y) as solution to the initial value problem

$$\frac{\partial u}{\partial y} = f_2(x,y,u),\qquad u(x,0) = \phi(x). \tag{7.6}$$

Hence it follows that u = u(x,y) is already uniquely defined by the specialized equation (7.5) and the second equation (7.3). The existence and uniqueness of u = u(x,y) follows from well-known theorems on ordinary differential equations. The constructed solution satisfies the first differential equation (7.3) at least at points of the real axis, but in general the first differential equation (7.3) is not everywhere satisfied. We are entitled to denote the system (7.3) as <u>overdetermined</u> because the function u = u(x,y) is already uniquely determined if the first equation (7.3) is replaced by the specialized one (7.5).

In the following we look for sufficient conditions under which the function u = u(x,y) constructed as solution of the initial value problem (7.6) turns out to satisfy the first equation (7.3) everywhere. For this end define the function U by

$$U(x,y) = \frac{\partial u}{\partial x}(x,y) - f_1(x,y,u(x,y)). \tag{7.7}$$

Since u(x,0) = $\phi(x)$ and since ϕ satisfies the specialized differential equation (7.5) one gets

$$U(x,0) = 0. \tag{7.8}$$

The right-hand side of the (ordinary) differential equation (7.6) is continuously differentiable. Applying well-known theorems on ordinary differential equations depending on parameters, we see that the second order derivative

$$\frac{\partial^2 u}{\partial y \partial x}$$

exists and, consequently, may be represented by

$$\frac{\partial^2 u}{\partial y \partial x} = \frac{\partial f_2}{\partial x} + \frac{\partial f_2}{\partial u}\frac{\partial u}{\partial x}. \tag{7.9}$$

From the definition (7.7) of U we obtain, moreover,

$$\frac{\partial U}{y} = \frac{\partial^2 u}{\partial x \partial y} - \frac{\partial f_1}{\partial y} - \frac{\partial f_1}{\partial u}\frac{\partial u}{\partial y}. \tag{7.10}$$

The partial derivative $\partial u/\partial y$ in the last formula may be replaced by the right-hand side f_2 of the second equation (7.3), whereas in view of the definition (7.7) the derivative $\partial u/\partial x$ in (7.9) may be replaced by

$(U + f_1)$. Taking into account the commutativity of the operations of partial differentiations, the formulae (7.10) and (7.9) yield

$$\frac{\partial U}{\partial y} = \frac{\partial f_2}{\partial x} + \frac{\partial f_2}{\partial u}(U + f_1) - \frac{\partial f_1}{\partial y} - \frac{\partial f_1}{\partial u}f_2. \tag{7.11}$$

In this expression the variables of the f_j and of their derivatives are x, y, and u(x,y). Now regard the point (x, y, u), where (x, y) is an arbitrary point belonging to G and u is an arbitrarily chosen real number. For an arbitrary point of this type define the function $\lambda = \lambda(x,y,u)$ by

$$\lambda = \frac{\partial f_2}{\partial x} - \frac{\partial f_1}{\partial y} + \frac{\partial f_2}{\partial u}f_1 - \frac{\partial f_1}{\partial u}f_2. \tag{7.12}$$

The overdetermined system (7.3) is said to be <u>completely integrable</u> if the function λ is equal to zero at each point (x, y, u). In this case $\lambda = \lambda(x,y,u)$ is also identically equal to zero if we substitute any function $u = u(x,y)$ depending on x and y instead of the third variable u. Then the same is true if $u = u(x,y)$ is the solution of the initial value problem (7.6). Therefore, in the case of completely integrable systems (7.3) the differential equation (7.11) passes into

$$\frac{\partial U}{\partial y} = \frac{\partial f_2}{\partial u}U.$$

It may be added that here the variables of $\partial f_2/\partial u$ are (x, y, u(x,y)). For each fixed x we interpret the last equation as ordinary differential equation (with respect to y). On the other hand, in view of (7.8) the initial values of U(x,y) are equal to zero at the point $y = 0$ for each x. Thus the function $U = U(x,y)$ must be equal to zero at each point (x,y) belonging to G. Consequently, the definition (7.7) of the function U implies that the function u constructed by solving the initial value problem (7.6) satisfies the first equation (7.3) everywhere in G, too.

Summarizing these considerations, the following lemma has been proved:

<u>Lemma.</u> Let G be a convex domain in the x,y-plane containing the origin. If the system (7.3) is completely integrable, then there exists a unique solution that is globally defined in G and satisfies the initial condition (7.4).

We would like to complete the lemma by some remarks. First let us hint at an analogous statement in the case that the right-hand sides of the system (7.3) depend not only on the independent variables and the desired function but also on derivatives of the sought function with respect to further (parametric) variables. The conditions under which such a system is completely integrable are very similar to those in the case of the system (7.3). Also in this case the desired solution can be constructed by freezing one independent variable. However, in order to guarantee the existence of solutions, further assumptions are to impose upon the right-hand sides (cf. the book [27] of E. Goursat; see also section 4.5. and the corresponding remark 1. on differential equations of type (0.3) in the introduction).

Next it should be mentioned that the lemma holds also in the case of more than two independent variables. The lemma is also true in the case of systems for several desired functions $u_1, \ldots, u_m$. Let us additionally remark that the concept of completely integrable systems can be generalized also to the case of systems in several complex variables (see [62]).

Finally we would like to hint to W. Rüprich's general theory [57] on systems of first order differential equations in both several real and several complex variables in which boundary value and initial value problems as well as problems with interior jump conditions are reduced to operator equations. Such problems (including their combinations) are solvable if the corresponding operator equations are completely integrable (the theory contains, especially, the concept of complete integrability for the corresponding operator equations).

Concluding our considerations on the first order system (7.3), we would like to discuss a special case which will be applied in section 7.4. dealing with an inverse problem for associated operators. We assume that the right-hand sides of (7.3) depend linearly on u, i.e., we specialize the system (7.3) to

$$\frac{\partial u}{\partial x} = \alpha_1 + \beta_1 u, \qquad \frac{\partial u}{\partial y} = \alpha_2 + \beta_2 u, \tag{7.13}$$

where the coefficients α_j, β_j depend continuously differentiable on x and y. Substituting $f_j = \alpha_j + \beta_j u$ into the right-hand side of (7.12), we get the expression

$$\left(\frac{\partial \alpha_2}{\partial x} - \frac{\partial \alpha_1}{\partial y} + \alpha_1\beta_2 - \alpha_2\beta_1\right) + \left(\frac{\partial \beta_2}{\partial x} - \frac{\partial \beta_1}{\partial y}\right)u.$$

The system (7.13) is completely integrable if the last expression equals to zero at each point (x, y, u). Thus the following statement is true:

The linear overdetermined system (7.13) is completely integrable provided both equations

$$\frac{\partial \alpha_2}{\partial x} - \frac{\partial \alpha_1}{\partial y} + \alpha_1\beta_2 - \alpha_2\beta_1 = 0. \tag{7.14}$$

$$\frac{\partial \alpha_2}{\partial x} - \frac{\partial \beta_1}{\partial y} = 0 \tag{7.15}$$

hold at each point of G.

7.3. An inverse problem for associated differential operators

Again let l and L be differential operators defined by

$$lw = \frac{\partial w}{\partial \bar{z}} - a(z)w - b(z)\bar{w},$$

$$Lw = C(z)\frac{\partial w}{\partial z} + A(z)w + B(z)\bar{w}.$$

Starting from the assumption that l is given, in section 6.4.3. operators L associated to l have been constructed. Conversely assume now that L is given. Then we look for operators l such that the given operator L is associated to the desired operator l. In order to be in a position to carry out the following calculations, we suppose that B and the real part of C are different from zero everywhere and, moreover, that A is twice continuously differentiable, while the derivatives of B up to the third order exist and are continuous. According to the theorem of section 6.4.2., the coefficient C is supposed to be holomorphic. Now we look for coefficients a(z), b(z) satisfying the relations (6.63), (6.64), and (6.65). From (6.63) it follows that b is uniquely determined,

$$b = \frac{B}{\bar{C}}. \tag{7.16}$$

Substituting (7.16) into (6.65), this relation passes into

$$\bar{C}a - C\bar{a} = A - \bar{A} + \frac{1}{B}\left(C\frac{\partial B}{\partial z} + \bar{C}\frac{\partial B}{\partial \bar{z}}\right) \tag{7.17}$$

because

$$\frac{\partial \bar{C}}{\partial z} = \overline{\left(\frac{\partial C}{\partial \bar{z}}\right)} = 0$$

and

$$\frac{\partial \bar{C}}{\partial \bar{z}} = \overline{\left(\frac{\partial C}{\partial z}\right)} = \overline{\left(\frac{dC}{dz}\right)}$$

(cf. sections 6.1.5. and 6.1.2.). The relation (6.64) and equation (7.17) show that the coefficient a(z) turns out to be a solution $\mu = \mu(z)$ of a system of type

$$\frac{\partial \mu}{\partial z} = \rho, \tag{7.18}$$

$$\bar{\lambda}\mu - \lambda\bar{\mu} = \kappa, \tag{7.19}$$

where

$$\rho = -\frac{1}{C}\frac{\partial A}{\partial \bar{z}}, \tag{7.20}$$

$$\lambda = C, \tag{7.21}$$

$$\kappa = A - \bar{A} + \frac{1}{B}\left(C\frac{\partial B}{\partial z} + \bar{C}\frac{\partial B}{\partial \bar{z}}\right). \tag{7.22}$$

Now we are going to reduce the system (7.18) and (7.19), consisting of a first order differential equation and a linear algebraic side condition, to an overdetermined system of type (7.13). For this end we split up both equations (7.18) and (7.19) into their real and imaginary parts. Using the denotations $\mu = \mu_1 + i\mu_2$, $\rho = \rho_1 + i\rho_2$, $\lambda = \lambda_1 + i\lambda_2$, $\kappa = \kappa_1 + i\kappa_2$, one gets

$$\frac{\partial\mu_1}{\partial x} + \frac{\partial\mu_2}{\partial y} = 2\rho_1, \tag{7.23}$$

$$\frac{\partial\mu_2}{\partial x} - \frac{\partial\mu_1}{\partial y} = 2\rho_2, \tag{7.24}$$

$$\kappa_1 = 0, \tag{7.25}$$

$$\lambda_1\mu_2 - \lambda_2\mu_1 = \frac{1}{2}\kappa_2. \tag{7.26}$$

The condition (7.25) is necessarily satisfied if the system (7.18), (7.19) possesses at least one solution. The equation (7.26) allows us to express μ_2 by μ_1:

$$\mu_2 = \sigma + \tau\mu_1, \tag{7.27}$$

where

$$\sigma = \frac{\kappa_2}{2\lambda_1}, \qquad \tau = \frac{\lambda_2}{\lambda_1}, \tag{7.28}$$

$\lambda_1 \neq 0$ everywhere. Substituting (7.27) into (7.23) and (7.24), we obtain the system

$$\begin{aligned} \frac{\partial\mu_1}{\partial x} &= \alpha_1 + \beta_1\mu_1, \\ \frac{\partial\mu_1}{\partial y} &= \alpha_2 + \beta_2\mu_1, \end{aligned} \tag{7.29}$$

where

$$\alpha_1 = \frac{1}{1+\tau^2}\left[(2\rho_1 - \frac{\partial\sigma}{\partial y}) + \tau(2\rho_2 - \frac{\partial\sigma}{\partial x})\right],$$

$$\alpha_2 = \frac{1}{1+\tau^2}\left[\tau(2\rho_1 - \frac{\partial\sigma}{\partial y}) - (2\rho_2 - \frac{\partial\sigma}{\partial x})\right],$$

$$\beta_1 = \frac{1}{1+\tau^2}(\frac{\partial\tau}{\partial y} + \tau\frac{\partial\tau}{\partial x}),$$

$$\beta_2 = \frac{1}{1+\tau^2}(\tau\frac{\partial\tau}{\partial y} - \frac{\partial\tau}{\partial x}).$$

The system (7.29) is an overdetermined system of type (7.13). In order to decide whether the system (7.29) is completely integrable or not, we calculate the left-hand sides of the equations (7.14) and (7.15). From the above definitions of α_1, α_2, β_1, β_2 one obtains easily

$$\frac{\partial\alpha_2}{\partial x} - \frac{\partial\alpha_1}{\partial y} + \alpha_1\beta_2 - \alpha_2\beta_1$$

$$= \frac{1}{(1+\tau^2)^2}\left[(2\rho_1 - \frac{\partial\sigma}{\partial y})(\frac{\partial\tau}{\partial x} + \tau\frac{\partial\tau}{\partial y}) + (2\rho_2 - \frac{\partial\sigma}{\partial x})(\tau\frac{\partial\tau}{\partial x} - \frac{\partial\tau}{\partial y})\right] \tag{7.30}$$

$$+ \frac{1}{1+\tau^2}\left[\frac{1}{2}\Delta\sigma + \tau\left(\frac{\partial\rho_1}{\partial x} - \frac{\partial\rho_2}{\partial y}\right) - \left(\frac{\partial\rho_2}{\partial x} + \frac{\partial\rho_1}{\partial y}\right)\right]$$

and

$$\frac{\partial\beta_2}{\partial x} - \frac{\partial\beta_1}{\partial y} = - \frac{2\tau}{(1+\tau^2)^2}\left\{\left(\frac{\partial\tau}{\partial x}\right)^2 + \left(\frac{\partial\tau}{\partial y}\right)^2\right\} + \frac{\Delta\tau}{1+\tau^2} . \tag{7.31}$$

Provided λ is holomorphic the right-hand sides of (7.30) and (7.31) may be simplified in the following way. Starting from the definitions (6.3) and (6.4) of the partial complex differentiations, we first note that

$$\frac{\partial}{\partial x} = \frac{\partial}{\partial z} + \frac{\partial}{\partial \bar{z}} ,$$

$$\frac{\partial}{\partial y} = i\left(\frac{\partial}{\partial z} - \frac{\partial}{\partial \bar{z}}\right).$$

Additionally taking into consideration the rules 6.1.5. and 6.1.2., it follows

$$\frac{\partial\lambda}{\partial x} = \frac{\partial\lambda}{\partial z} + \frac{\partial\lambda}{\partial\bar{z}} = \frac{d\lambda}{dz} ,$$

$$\frac{\partial\bar{\lambda}}{\partial x} = \frac{\partial\bar{\lambda}}{\partial z} + \frac{\partial\bar{\lambda}}{\partial\bar{z}} = \overline{\left(\frac{\partial\lambda}{\partial\bar{z}}\right)} + \overline{\left(\frac{\partial\lambda}{\partial z}\right)} = \overline{\left(\frac{d\lambda}{dz}\right)}$$

and analogously,

$$\frac{\partial\lambda}{\partial y} = i\frac{d\lambda}{dz}, \qquad \frac{\partial\bar{\lambda}}{\partial y} = - i\overline{\left(\frac{d\lambda}{dz}\right)}.$$

In view of $\lambda = \lambda_1 + i\lambda_2$, $\bar{\lambda} = \lambda_1 - i\lambda_2$, $\lambda_1 = \frac{1}{2}(\lambda + \bar{\lambda})$, $\lambda_2 = \frac{i}{2}(\bar{\lambda} - \lambda)$ one has

$$\tau = \frac{\lambda_2}{\lambda_1} = i\frac{\bar{\lambda} - \lambda}{\lambda + \bar{\lambda}} ,$$

and the last formulae yield

$$\frac{\partial\tau}{\partial x} = \frac{2i}{(\lambda + \bar{\lambda})^2}\left\{C\overline{\left(\frac{dC}{dz}\right)} - \bar{C}\frac{dC}{dz}\right\}$$

and similar expressions for $\frac{\partial\tau}{\partial y}$, $\frac{\partial\tau}{\partial x} + \tau\frac{\partial\tau}{\partial y}$ and so on. Using these expressions, we see that the right side of (7.31) is equal to zero. Therefore, the condition (7.15) is always satisfied provided λ is holomorphic.

Second, the definitions (6.3) and (6.4) lead to representations of the terms on the right side of (7.30) by complex quantities, e.g. we have

$$\frac{\partial\rho}{\partial\bar{z}} = \frac{1}{2}\left(\frac{\partial\rho_1}{\partial x} - \frac{\partial\rho_2}{\partial y}\right) + \frac{i}{2}\left(\frac{\partial\rho_2}{\partial x} + \frac{\partial\rho_1}{\partial y}\right).$$

Hence it follows

$$\frac{\partial\rho_1}{\partial x} - \frac{\partial\rho_2}{\partial y} = \frac{\partial\rho}{\partial\bar{z}} + \overline{\left(\frac{\partial\rho}{\partial\bar{z}}\right)}$$

and so on. In this way the condition (7.14) may be rewritten as

$$\overline{\left(\frac{d\lambda}{dz}\right)}\left(i\rho + \frac{\partial\sigma}{\partial z}\right) + \frac{d\lambda}{dz}\left(- i\rho + \frac{\partial\sigma}{\partial\bar{z}}\right) \tag{7.32}$$

$$+ (\lambda + \overline{\lambda}) \frac{\partial^2\sigma}{\partial z \partial \overline{z}} + i\left\{\overline{\lambda}\frac{\partial \rho}{\partial \overline{z}} - \lambda\overline{\left[\frac{\partial \rho}{\partial \overline{z}}\right]}\right\} = 0.$$

It is clear that (7.32) is one real equation because the imaginary part of the left-hand side is identically equal to zero. Expressing the terms in (7.32) by their real and imaginary parts, the last condition (7.32) may also be written in the form

$$\mathrm{Re}\left[\overline{\left[\frac{d\lambda}{dz}\right]}(i\rho + \frac{\partial\sigma}{\partial z})\right] + \mathrm{Re}\ \lambda\cdot\frac{\partial^2\sigma}{\partial z \partial \overline{z}} - \mathrm{Im}\left[\overline{\lambda}\frac{\partial\rho}{\partial \overline{z}}\right] = 0. \tag{7.33}$$

Summarizing these calculations, the following result has been proved:

If λ is holomorphic and if the condition (7.32) is satisfied, then the system (7.29) is completely integrable. Then (by virtue of the lemma of 7.2.) there exists a solution μ_1 to (7.29) (in convex domains) taking an arbitrarily prescribed value at one point.

Define μ_2 by (7.27). Then (μ_1, μ_2) is proved to be a solution to (7.23), (7.24). Thus the complex-valued function $\mu = \mu(z)$ is a solution to the complex differential equation (7.18). The complex side condition (7.19) is satisfied, too, provided the real part κ_1 of the right-hand side κ equals identically to zero. The above-mentioned result proves the following lemma:

<u>Lemma.</u> Let G be a convex domain in the z-plane. If the conditions (7.25) and (7.32) are satisfied, then there exists a uniquely determined solution to the system (7.18), (7.19) whose real part may be chosen arbitrarily at one point of G.

Now return to the original inverse problem for associated operators. We reduced this problem to the complex differential equation (7.18) with the complex side condition (7.19), where the coefficients ρ, λ, and κ are defined by (7.20), (7.21), and (7.22), respectively. Since the real part of $A - \overline{A}$ is equal to zero, in view of (7.22) the condition (7.25) may be rewritten in the form

$$\mathrm{Re}\left[\frac{1}{B}(C\frac{\partial B}{\partial z} + \overline{C}\frac{\partial B}{\partial \overline{z}})\right] = 0. \tag{7.34}$$

On the other hand, assumption (7.25) implies that the first equation (7.28) may be replaced by

$$\sigma = - i\frac{\kappa}{\lambda + \overline{\lambda}} . \tag{7.35}$$

Therefore in view of (7.21) and (7.22) the function σ may be represented by

$$\sigma = - \frac{i}{C + \overline{C}}\left\{A - \overline{A} + \frac{1}{B}(C\frac{\partial B}{\partial z} + \overline{C}\frac{\partial B}{\partial \overline{z}})\right\}$$

provided the condition (7.34) is satisfied. Additionally taking into consideration equation (7.20), condition (7.33) may be written as

$$\mathrm{Re}\left[\left(\overline{\frac{dC}{dz}}\right)\left(-\frac{i}{C}\frac{\partial A}{\partial \bar z}+\frac{\partial \sigma}{\partial z}\right)\right]+\frac{1}{2}(C+\bar C)\frac{\partial^2\sigma}{\partial z\partial \bar z}+\mathrm{Im}\left[\frac{\bar C}{C}\frac{\partial^2 A}{\partial \bar z^2}\right]=0, \qquad (7.36)$$

where σ is given by (7.35).

Concerning the problem formulated at the beginning of this chapter, we have proved the following theorem:

<u>Theorem.</u> Let G be a convex domain. Suppose that the coefficients $A(z)$, $B(z)$, $C(z)$ of L satisfy the conditions (7.34) and (7.36), where C is holomorphic in G. Then there exist coefficients $a(z)$, $b(z)$ such that L is associated to the operator l with the coefficients $a(z)$, $b(z)$. The real part of $a(z)$ may be arbitrarily chosen at one point of G.

Note that (7.34) is a first order differential equation, whereas (7.36) is a third order equation. The equation (7.36) contains the term

$$\frac{\partial}{\partial x}\Delta B$$

if $C(z) = 1$ everywhere, whereas it is a second order equation for A howsoever the holomorphic function $C = C(z)$ is chosen.

If the coefficients of L depend on t additionally, i.e., if we have

$$A = A(z,t), \quad B = B(z,t), \quad C = C(z,t),$$

then the coefficients of l constructed by the method explained above do also depend on t, in general. For the purpose of applications of the concept of associated differential operators in the theory of initial value problems (see the following section 7.4.), however, the coefficients of l must not depend on t. In order to ensure such independence, further conditions must be imposed upon the given coefficients $A(z,t)$, $B(z,t)$, $C(z,t)$. E.g., the coefficients $a(z)$, $b(z)$ constructed above are independent of t if

$$\frac{B}{\bar C}, \quad \frac{1}{C}\frac{\partial A}{\partial \bar z}, \quad \sigma, \quad \text{and} \quad \tau$$

do not depend on t.

Conversely, analogous considerations may be carried out if the coefficients of L are given. Similar constructions (in both cases) are possible also under the assumption that the coefficients belong to an L_p-space.

Let us additionally also hint to section 6.4.4. in which we constructed associated operators in the case that the given coefficients as well as the desired ones are piecewise constant.

7.4. Construction of solutions with prescribed generalized analytic initial functions

Regard the initial value problem

$$\frac{\partial w}{\partial t} = C(t,z)\frac{\partial w}{\partial z} + A(t,z)w + B(t,z)\overline{w} + D(t,z),$$
$$w(0,z) = w_0(z), \tag{7.37}$$

where the desired function $w = w(t,z)$ is complex-valued and depends on both the real variable t and the complex one z. Denote the differential operator defined by the right-hand side of the first equation (7.37) by $\tilde{L}w$ (cf. section 7.1.). Further denote

$$C(t,z)\frac{\partial w}{\partial z} + A(t,z)w + B(t,z)\overline{w}$$

by Lw. Therefore, the operator L may be interpreted as homogeneous part of $\tilde{L}$, and L and $\tilde{L}$ are connected by the relation

$$\tilde{L}w = Lw + D(t,z).$$

Let G be a given bounded domain in the z-plane and let, moreover, T be a given positive number. Then suppose that the coefficients $A(t,z)$, $B(t,z)$, $C(t,z)$, and $D(t,z)$ satisfy the following conditions:

a) They are continuous in

$$\{t : 0 \leq t \leq T\} \times \overline{G},$$

and for every t Hölder-continuous in $\overline{G}$ with a fixed Hölder-exponent α, $0 < \alpha < 1$. For each t the Hölder-norms of the coefficients are supposed to be uniformly bounded.

b) There exists a differential operator l of type (6.59),

$$lw = \frac{\partial w}{\partial \overline{z}} - a(z)w - b(z)\overline{w},$$

such that L, the homogeneous part of $\tilde{L}$, is associated to l. The coefficients $a(z)$ and $b(z)$ of l do not depend on t and are supposed to be Hölder-continuous with the same exponent α in $\overline{G}$.

c) Suppose that the absolute term $D(z,t)$ of the differential operator $\tilde{L}$ satisfies the differential equation

$$lD(t,\cdot) = 0$$

for each t.

Remark 1. Using the method of section 7.3., we may construct a differential operator l such that the given operator L is associated to l. In order to apply this method, the coefficients $A(t,z)$, $B(t,z)$, $C(t,z)$, and $D(t,z)$ must satisfy differentiability conditions stronger than a).

Remark 2. Both conditions b) and c) imply that $\tilde{L}$ is associated to l, too.

The three conditions a), b), and c) on the operator L are completed by the following two conditions on the initial function:

a) The initial function is Hölder-continuous with the exponent α in $\overline{G}$.

b) Further, the initial function satisfies the differential equation

$$lw_0 = 0$$

in G.

Now we are going to construct a solution to the initial value problem (7.37) provided the above-mentioned assumptions on the differential operator L and the initial function $w_0 = w_0(z)$ are satisfied. The solution will be constructed by using successive approximations in scales of Banach spaces. For this end we choose a family of subdomains G_s, $0 < s < s_0$, of the given domain such that the corresponding conditions a), b), and c) formulated in section 2.2. are satisfied. Additionally assume that the norms of the T_{G_s}- and Π_{G_s}-operators (cf. 6.2.7. and 6.2.14. resp.) belonging to the subdomains G_s are uniformly bounded [1]) (the norm of the T_G-operator tends to zero as the diameter of G tends to zero, cf., for instance, the proof of theorem 2 in section 12.1. of [64]; in view of lemma 7 on p. 169 of the same book [64], the norms of the Π_G-operators are uniformly bounded provided the domains G are disks centred at the same point). Again denote the constant in the condition (2.5) by c, i.e.,

$$\text{dist}\,(G_{s'}, \partial G_s) \geq c(s - s') \tag{7.38}$$

if $0 < s' < s < s_0$ (see (4.25) in section 4.3.). Carrying out the limiting process $s \to s_0$ and after that rewriting s' as s, from (7.38) one obtains the estimate

$$\text{dist}\,(G_s, \partial G) \geq c(s_0 - s), \tag{7.39}$$

too.

Let W_s be the space of all solutions of the (associated) differential equation $lw = 0$ which are Hölder-continuous in $\overline{G}_s$. By virtue of 6.3.8. each W_s turns out to be a Banach space equipped with the norm defined in 6.2.6. For short denote the norm in $C^\alpha(\overline{G}_s)$ by $\|\cdot\|_s$. Take any function $w = w(z)$ belonging to W_s. Resticting it to the subset $G_{s'}$, $s' < s$, we see that $W_{s'}$ is injected into W_s (cf. section 2.2.). Thus the family W_s of Banach spaces, $0 < s < s_0$, forms a scale. In order to solve the initial value problem (7.37) by applying the theorem of section 3.8.,

1) In his thesis [16] A. Crodel avoids this assumption by extending a Hölder-continuous function in $\overline{G}_s$ to an analogous function in $\overline{G}$ (such extensions are constructed, for instance, in E.J.McShane's paper [38]).

we have to show that the right-hand side of the differential equation (7.37) may be interpreted as operator mapping the scale W_s into itself and satisfying the necessary conditions. In view of 6.3.7. the derivative $\partial w/\partial \bar{z}$ is Hölder-continuous (with the same exponent) in $\bar{G}_{s'}$, provided $w = w(z)$ belongs to W_s, $s > s'$. Furthermore the coefficients of the differential operator on the right-hand side of the first equation (7.37) depend Hölder-continuously on z. Thus the right-hand side of the first equation (7.37) defines an operator mapping W_s into $W_{s'}$. Now substitute a function $w = w(t,z)$ depending on both variables t and z into the right-hand side of the first equation (7.37). Since the coefficients depend continuously on t the right-hand side is a continuous function in t provided $w(t,z)$ depends continuously on t. Therefore all successive approximations (3.19) exist. This means that condition (I') need not be necessarily satisfied (see also the corresponding remark after the theorem in section 3.8.; it may be added that condition (I') is satisfied, for instance if the Hölder-norms of the coefficients depend continuously on t).

It remains to verify that the conditions (II) and (III) of section 3.2. are satisfied in the case of the operator defined by the right-hand side of the first equation (7.37). Substituting the given initial function $w_0 = w_0(z)$ into this right-hand side, one gets

$$C(t,z)\frac{\partial w_0}{\partial z} + A(t,z)w_0 + B(t,z)\bar{w}_0 + D(t,z). \tag{7.40}$$

According condition a) on the initial function, the norm $\|w_0\|_{C^\alpha(\bar{G})}$ is finite. Since the product of two Hölder-continuous functions is again Hölder-continuous we have (see also 6.3.9.)

$$\|A(t,\cdot)w_0\|_s \leq \|A(t,\cdot)w_0\|_{C^\alpha(\bar{G})} \leq 2\sup_{0\leq t\leq T}\|A(t,\cdot)\|_{C^\alpha(\bar{G})}\|w_0\|_{C^\alpha(\bar{G})}.$$

An analogous estimate holds for $B(t,z)\bar{w}_0$. Taking into account the estimate (7.39), by virtue of 6.3.9. we obtain an estimate of type

$$\left\|\frac{\partial w_0}{\partial z}\right\|_s \leq \frac{\text{const}}{s_0 - s}\|w_0\|_{C^\alpha(\bar{G})}.$$

Thus the s-norm of the first term in (7.40) may be estimated by

$$2\|C(t,\cdot)\|_s\cdot\frac{\text{const}}{s_0 - s}\|w_0\|_{C^\alpha(\bar{G})}$$

$$\leq 2\sup_{0\leq t\leq T}\|C(t,\cdot)\|_{C^\alpha(\bar{G})}\cdot\frac{\text{const}}{s_0 - s}\|w_0\|_{C^\alpha(\bar{G})}.$$

Summarizing these estimates, we see that the s-norm of the whole expression (7.40) may be estimated by a quantity of the form

$$\frac{K_1}{s_0 - s} + K_2, \tag{7.41}$$

where K_1 and K_2 are certain constants depending on $\|w_0\|_{C^\alpha(\overline{G})}$ and on the bounds for the Hölder-norms of the coefficients $A(t,z)$, $B(t,z)$, $C(t,z)$, and $D(t,z)$ that exist in accordance with our assumptions. Since

$$s_0 - s < s_0$$

implies

$$1 < \frac{s_0}{s_0 - s},$$

the expression (7.41) can be estimated by

$$\frac{K_1}{s_0 - s} + K_2 \leq \frac{K_1 + s_0 K_2}{s_0 - s}$$

and, consequently, the condition (II) is satisfied, where

$$K = K_1 + s_0 K_2.$$

In order to prove that condition (III) is satisfied, too, we substitute two elements w_1 and w_2 of W_s into the right-hand side of the first equation (7.37). The difference of the arising expressions equals to

$$C(t,z)\left(\frac{\partial w_2}{\partial z} - \frac{\partial w_1}{\partial z}\right) + A(t,z)(w_2 - w_1) + B(t,z)(\overline{w}_2 - \overline{w}_1). \qquad (7.42)$$

The s-norm and, consequently, also the s'-norm of the second and the third term can be estimated by a quantity of type

$$\text{const}\|w_2 - w_1\|_s.$$

We get this estimate in a similar way as we estimated the second and the third term in (7.40). In order to estimate the first term of (7.42), we make use of 6.3.9., again. Taking into consideration the estimate (7.38) of the distance of $G_{s'}$ from the boundary of G_s, we obtain that the s'-norm of the first term in (7.42) can be estimated by an expression of the form

$$\frac{\text{const}}{s - s'}\|w_2 - w_1\|_s.$$

Summarizing these inequalities, it follows that the s'-norm of (7.42) is not larger than an expression of type

$$\left(\text{const} + \frac{\text{const}}{s - s'}\right)\|w_2 - w_1\|_s.$$

Since $s - s' < s < s_0$ implies

$$1 < \frac{s_0}{s - s'}$$

the last bound can be replaced by a bound of the form

$$\frac{C}{s - s'}\|w_2 - w_1\|_s. \qquad (7.43)$$

This means that condition (III) of section 3.2. is satisfied, too. Applying the theorem of section 3.8., we have proved, consequently, the following theorem (cf. [66]):

Theorem. Suppose that there exist a differential operator l,

$$lw = \frac{\partial w}{\partial \bar{z}} - a(z)w - b(z)\bar{w},$$

such that the differential operator

$$Lw = C(t,z)\frac{\partial w}{\partial z} + A(t,z)w + B(t,z)\bar{w}$$

is associated to l. Suppose, moreover, that the absolute term $D(t,z)$ as well as the initial function $w_0 = w_0(z)$ are solutions of the associated differential equation $lw = 0$. Provided the coefficients $A(t,z)$, $B(t,z)$, $C(t,z)$, and $D(t,z)$ are continuous in t and Hölder-continuous in z the initial value problem (7.37) is solvable by a function $w = w(t,z)$ satisfying the associated differential equation

$$lw(t,\cdot) = 0$$

at each t. The solution exists in G_s and belongs to W_s if

$$0 \leq t < \frac{1}{Ce}(s_0 - s),$$

where C is the constant occuring in (7.43). The solution can be constructed by successive approximations:

$$w_{k+1}(t,z) = w_0(z) + \int_0^t (\tilde{L}w_k)(\tau,z)d\tau.$$

$k = 0, 1, 2, \ldots$ ($w_0(\tau,z)$ is to be replaced by the values $w_0(z)$ of the initial function).

Let us add that the approximations $w_k = w_k(t,z)$ converge with respect to the Hölder-norm 6.2.6. Let us additionally also emphasize that the definition of w_{k+1} contains the operator $\tilde{L}$. The calculation of w_{k+2} requires once more to apply the operator $\tilde{L}$. In view of the theorem in section 6.5. the operator L and, consequently, also the operator $\tilde{L}$ may be repeatedly applied to solutions of the associated differential equation $lw = 0$.

Finally, it should be mentioned that there is the following link between the above theorem and H. Lewy's example [35]: The theorem shows that the initial value problem (7.37) is solvable provided that there exists an associated differential equation $lw = 0$. On the other hand, the H. Lewy example is a differential equation of the type of the first equation (7.37) that does not possess any solution. Thus it is natural that an associated differential equation need not exist in any case.

7.5. Uniqueness theorems for initial value problems with generalized analytic initial functions

The initial value problem (7.37) has been solved by reducing it to an initial value problem for operators acting in a suitably chosen scale of Banach spaces. The solution of such abstract initial value problems is uniquely determined in the scale (see the theorem in section 3.7. and also the theorem in section 3.8.). The unique solvability in the scale does not exclude, however, the existence of further solutions not belonging to the chosen scale. In the holomorphic case (cf. chapter 4) the non-existence of further solutions, not belonging to the corresponding scale of Banach spaces of holomorphic functions, had been excluded by the Holmgren theorem (cf. Chapter 5). The immediate application of this theorem yields also uniqueness theorems for the initial value problem (7.37). These uniqueness theorems exclude the existence of solutions of the initial value problem (7.37) that do not belong to the scale of Banach spaces W_s of generalized analytic functions constructed in the preceding section 7.4. The application of the classical Holmgren theorem requires, however, that the coefficients A(t,z), B(t,z), and C(t,z), are representable by power series in t, z, and $\bar{z}$ or at least by deformed power series (cf. section 5.4.).

First assume that the coefficients under consideration are power series in t, z, and $\bar{z}$, i.e., they are power series in t, x, and y (we take into account that z, $\bar{z}$ may be expressed linearly by x and y and vice versa). The difference of two solutions to the same initial value problem (7.37) is a solution to the corresponding homogeneous equation (for which D(t,z) is equal to zero everywhere) and vanishes identically at t = 0. Applying the classical Holmgren theorem for homogeneous differential equations (see the sections 5.1. and 5.2.), it follows that the difference of the two solutions equals to zero everywhere. Thus the following theorem has been proved:

<u>Theorem 1.</u> The initial value problem (7.37) possesses at most one solution provided the coefficients A(t,z), B(t,z), and C(t,z) are representable by power series in t, z, and $\bar{z}$.

Now assume that the coefficients in question are deformed power series at the point z = 0 (cf. 5.4.), i.e., they are representable by power series in x and y with coefficients depending continuously on $t' = t + \lambda(x,y)$:

$$\sum_{\nu,\mu} c_{\nu\mu}(t')x^{\nu}y^{\mu}.$$

The last representation is equivalent to a representation of the form

$$\sum_{\nu,\mu} d_{\nu\mu}(t')z^{\nu}\bar{z}^{\mu}.$$

Applying the generalized Holmgren theorem (cf. section 5.4.) to the difference of two different solutions existing possibly, one gets readily the following uniqueness theorem:

Theorem 2. Assume that the coefficients $A(t,z)$, $B(t,z)$, and $C(t,z)$ are deformed power series with respect to $z = 0$. Then there exists a neigbourhood of the origin of the (t,z)-space in which the initial value problem (7.37) possesses at most one solution.

8. CONTRACTION-MAPPING PRINCIPLES IN SCALES OF BANACH SPACES

In chapter 3 we have explained a general method for solving initial value problems in abstract scales of Banach spaces. This abstract version of the Cauchy-Kovalevskaya theorem is advantageous because it comprises not only the case of initial value problems for differential equations (cf. the chapters 4 and 7) but also uniqueness theorems of the Holmgren type (cf. chapter 5, section 5.2., especially).

On the other hand, the use of time-depending norms allows us to reduce initial value problems to contraction-mapping principles without using the concept of scales of Banach spaces. In this way one gets, consequently, elementary constructive proofs for the Cauchy-Kovalevskaya theorems. For the classical Cauchy-Kovalevskaya theorem this was done by W. Walter who used a weighted maximum norm in his paper [78] (cf. section 8.1.). Since the Π_G-operator is not a bounded operator in the space of continuous functions equipped with the maximum norm this method cannot be immediately applied to the case of generalized analytic functions. Modifying W. Walter's definition of a time-depending norm, in section 8.2. an elementary proof of the Cauchy-Kovalevskaya theorem for generalized analytic functions will be given. Once more generalizing this construction, in 8.3. a contraction-mapping principle for initial value problems in abstract scales of Banach spaces will be given.

8.1. W. Walter's elementary proof of the classical Cauchy-Kovalevskaya theorem

The proof of the classical Cauchy-Kovalevskaya theorem carried out in section 4.3. is based on the method of successive approximations in a suitably chosen scale of Banach spaces. The norm in these spaces is the supremum norm (maximum norm) defined with respect to the spacelike variable z. W. Walter uses a modified supremum norm including also the variable t. This norm allows to prove the (classical) Cauchy-Kovalevskaya theorem with the help of Banach's fixed-point theorem in a fixed

Banach space, not in a scale of Banach spaces (see W. Walter's paper [78]).

The proof of the classical Cauchy-Kovalevskaya theorem carried out by using the method of scales of Banach spaces in chapter 4 is based on the estimate (2.2) of the derivative of a holomorphic function. Also W. Walter's proof is based on a slight modification of this estimate. This modification is given by Nagumo's lemma. Let G be a domain in the z-plane. First we define the distance function $d = d(z)$ for points z belonging to G by

$$d(z) = \inf_{\zeta\in\partial G}|z - \zeta|.$$

It is clear that $d = d(z)$ is a positive and continuous function in G approaching zero as z tends to the boundary of G.

Let, further, $w = w(z)$ be a holomorphic function defined in G and satisfying the inequality

$$|w(z)|\, d^p(z) \leq c \qquad (8.1)$$

everywhere in G, where c and p are given real constants, $p \geq 0$. The inequality (8.1) ensures that the possible growth of $w = w(z)$ near the boundary of G is restricted. Then the Nagumo lemma ([44]) states:

Lemma 1. Suppose the function $w = w(z)$ defined and holomorphic in G satisfies the inequality (8.1) everywhere in G. Then its derivative may be estimated by

$$\left|\frac{dw}{dz}(z)\right| d^{p+1}(z) \leq cC_p$$

for each z belonging to G, where

$$C_p = (1 + p)(1 + \frac{1}{p})^p, \quad p > 0,$$
$$C_0 = 1.$$

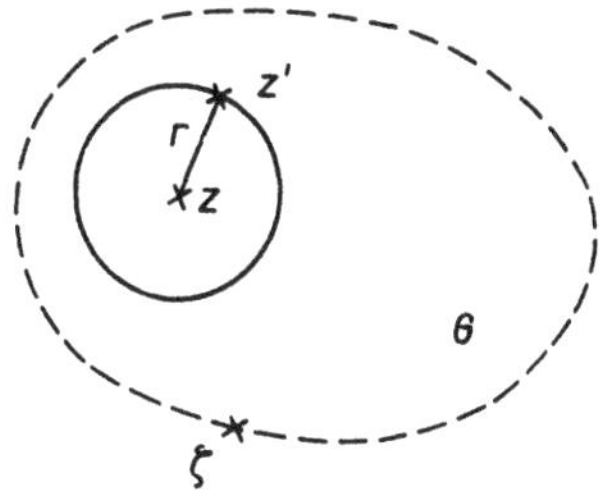

Proof. Let z be an arbitrary point of G. Take any positive number r with $r < d(z)$. Then the closed disk centred at z with radius r belongs to G. In view of Cauchy's integral formula we have

$$\frac{dw}{dz}(z) = \frac{1}{2\pi i}\int_{|z'-z|=r}\frac{w(z')}{(z'-z)^2}dz'. \qquad (8.2)$$

In view of the triangle inequality we have

$$|z - \zeta| \leq |z - z'| + |z' - \zeta| = r + |z' - \zeta|$$

for each point ζ of the boundary ∂G. Therefore one gets also

$$\inf_{\zeta\in\partial G}|z - \zeta| \leq r + \inf_{\zeta\in\partial G}|z' - \zeta|,$$

i.e.,

$$d(z) \leq r + d(z'). \tag{8.3}$$

The assumption (8.1) yields, especially,

$$|w(z')| \leq \frac{c}{d^p(z')} .$$

In view of (8.3) we obtain, consequently,

$$|w(z')| \leq \frac{c}{(d(z) - r)^p} .$$

Therefore, (8.2) implies

$$\left|\frac{dw}{dz}(z)\right| \leq \frac{1}{2\pi}\cdot\frac{1}{r^2}\cdot\frac{c}{(d(z) - r)^p}\cdot 2\pi r \leq \frac{1}{r}\,\frac{c}{(d(z) - r)^p} . \tag{8.4}$$

Substituting

$$r = \frac{d(z)}{1 + p} ,$$

we obtain

$$\left|\frac{dw}{dz}(z)\right| \leq \frac{c(1 + p)}{d^{p+1}(z)}\left(\frac{p + 1}{p}\right)^p$$

provided $p > 0$. This estimate, however, is identical with the statement of the asserted lemma in the case $p > 0$. In the case $p = 0$ the inequality (8.4) passes into

$$\left|\frac{dw}{dz}(z)\right| \leq \frac{c}{r} ,$$

and the limiting process $r \to d(z)$ shows that the lemma holds also in this case.

Notice that

$$C_p < (1 + p)e$$

for each $p \geq 0$. Furthermore we would like to remark that the lemma is true in the case of several complex variables, too. Assume that G is a domain in $\mathbb{C}^n$, $z = (z_1, \ldots, z_n)$. Applying the Cauchy integral formula with respect to the variable z_j, we get

$$\frac{\partial w}{\partial z_j}(z) = \frac{1}{2\pi i}\int_{|z_j'-z_j|=r}\frac{w(z')}{(z_j' - z_j)^2}dz_j'$$

instead of (8.2), where $z' = (z_1, \ldots, z_{j-1}, z_j', z_{j+1}, \ldots, z_n)$. In the case of several complex variables one obtains, therefore, the following statement:

Suppose that $w = w(z)$ is holomorphic in the domain G of $\mathbb{C}^n$ and satisfies the inequality (8.1). Then for each $j = 1, \ldots, n$ the complex derivative with respect to z_j may be estimated by

$$\left|\frac{\partial w}{\partial z_j}(z)\right| d^{p+1}(z) \leq cC_p.$$

Introduce now the Banach space $H_*(G)$ whose elements are holomorphic functions in G such that

$$|w(z)|\,d^p(z)$$

is bounded in G, where G is a given domain in the complex plane or in $\mathbb{C}^n$. It is clear thath $H_*(G)$ is a linear space. Further

$$\sup_G |w(z)|\,d^p(z) = \|w\|_* \tag{8.5}$$

defines a norm (a weighted maximum norm) in $H_*(G)$ because we have, first, that $\|w\|_* \geqq 0$, and $\|w\|_* = 0$ if and only if w vanishes identically in G. Second we see immediately that

$$\|\alpha w\|_* = |\alpha| \sup_G |w(z)|\,d^p(z) = |\alpha|\cdot\|w\|_*,$$

where α is a complex constant. It remains to prove that the norm $\|\cdot\|_*$ defined by (8.5) satisfies the triangle inequality. For this end we take two functions w_1, w_2 belonging to $H_*(G)$. Then we have

$$|w_1(z) + w_2(z)|\,d^p(z)$$

$$\leqq \sup_G |w_1(z)|\,d^p(z) + \sup_G |w_2(z)|\,d^p(z) = \|w_1\|_* + \|w_2\|_*$$

for each point for z and, consequently,

$$\|w_1 + w_2\|_* \leqq \|w_1\|_* + \|w_2\|_*.$$

Now let K be any compact subset of G whose distance from the boundary of G is at least equal to δ. Then we have

$$|d(z)| \geqq \delta$$

in K. Thus the definition (8.5) of $\|\cdot\|_*$ yields the estimate

$$|w(z)| \leqq \frac{1}{\delta^p}\|w\|_* \tag{8.6}$$

for each point belonging to K.

This estimate allows us to prove the completeness of $H_*(G)$. For this end we take any fundamental sequence $\{w_n\}_{n=1,2,\ldots}$ in $H_*(G)$, i.e.,

$$\|w_n - w_m\|_* < \varepsilon \tag{8.7}$$

for sufficiently large n and m. Applying (8.6) to $w_n - w_m$, we get

$$|w_n(z) - w_m(z)| \leqq \frac{\varepsilon}{\delta^p}$$

for each z of K. Hence the w_n are uniformly convergent on K. By virtue of Weierstrass' convergence theorem the limit function $w = w(z)$ is holomorphic in all (inner) points of K. Since K is an arbitrary compact subset of G the limit function exists everywhere in K and is holomorphic in G (for an arbitrary point z of G we choose K in such a manner that z is an inner point of K). Taking into account the definition (8.5) of $\|\cdot\|_*$, the condition (8.7) yields

$$|w_n(z) - w_m(z)|d^p(z) < \varepsilon$$

for each z of G. Carrying out the limiting process $m \to \infty$, the last inequality gives

$$|w_n(z) - w(z)|d^p(z) \leq \varepsilon \tag{8.8}$$

everywhere in G (we have $w_m(z) \to w(z)$ for $m \to \infty$ everywhere in G, although the convergence is not uniform, in general). The inequality (8.8) means that

$$\|w_n - w\|_* \leq \varepsilon \tag{8.9}$$

for sufficiently large n. On the other hand we have

$$|w(z)|d^p(z) \leq |w(z) - w_n(z)|d^p(z) + |w_n(z)|d^p(z) \leq \varepsilon + \|w_n\|_* .$$

Thus the limit function $w = w(z)$ itself belongs to $H_*(G)$. In view of (8.9) the w_n approaches w in the sense of the metric of $H_*(G)$. Summing up these considerations we have shown that the following theorem holds:

Theorem 1. The space $H_*(G)$ turns out to be a Banach space equipped with the modified supremum norm $\|\cdot\|_*$.

In addition we would like to remaind that the elements of $H_*(G)$ are complex-valued functions. Notice that an analogous statement is true if the elements w of $H_*(G)$ are vectors $w = (w_1, \dots, w_m)$ with complex-valued components defined in G. In this case the modified supremum norm is defined by

$$\|w\|_* = \max_j \|w_j\|_* .$$

In order to apply the above considerations for proving the Cauchy-Kovalevskaya theorem, we have to admit that the holomorphic functions in question depend on the time, in addition. First we have to describe the set in the t,z-space in which the functions $w = w(t,z)$ are defined. We start from a given domain G in the z-plane (or in $\mathbb{C}^n$). Define

$$M = \{(t,z) : z \in G, 0 \leq t < \eta d(z)\},$$

where $\eta > 0$ will be specified below. The set M is some kind of conical set with the base G (cf. the figure). Define, further,

$$d(t,z) := d(z) - \frac{t}{\eta}.$$

This function is positive in all inner points of the conical set M. It approaches zero as (t,z) approaches the lateral area.

Now define $H_*(M)$ as the space of all functions $w = w(t,z)$ defined and continuous in M and holomorphic in z (for fixed t) for which

$$|w(t,z)|\,d^p(t,z)$$

is bounded in M. Analogously to (8.5) the expression

$$\sup_M |w(t,z)|\,d^p(t,z) = \|w\|_* \qquad (8.10)$$

defines a norm in $H_*(M)$.

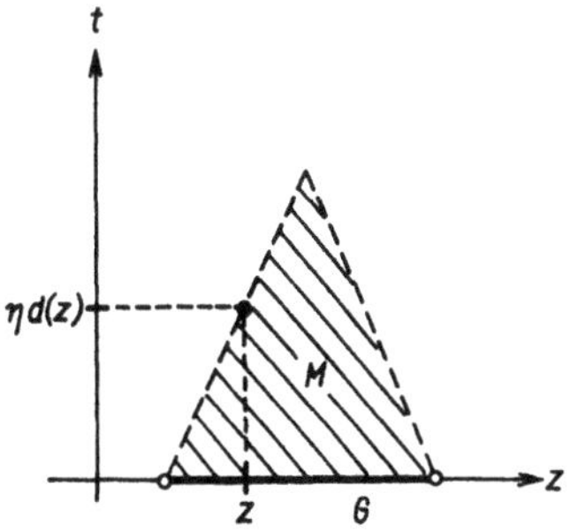

Now we are going to prove that $H_*(M)$ is a Banach space. For this end we introduce compact subsets of M having the property that the upper lower bound of d(t,z) is positive in these compact subsets. We start from any compact subset K of G. Denote the distance of K from the boundary ∂G of G by δ. Then define

$$\tilde{K} = \{(t,z) : z \in K,\ 0 \leqq t \leqq \eta(d(z) - \delta)\}. \qquad (8.11)$$

Therefore, for any point (t,z) belonging to $\tilde{K}$ we have

$$-\frac{t}{\eta} \geqq -d(z) + \delta$$

and, consequently,

$$d(t,z) \geqq \delta. \qquad (8.12)$$

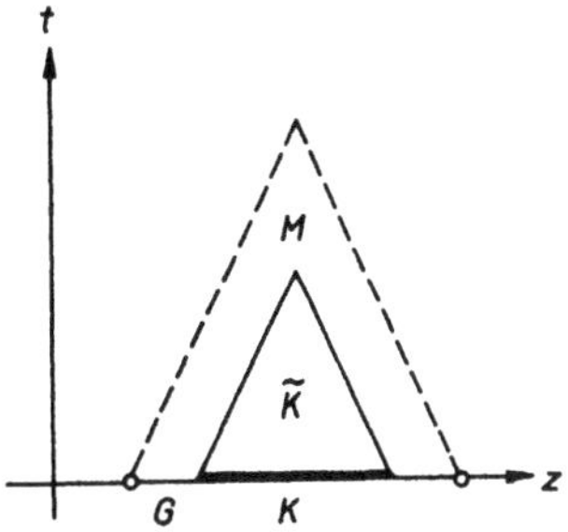

Now let (t',z') be any point of M. Hence we have

$$z' \in G, \quad 0 \leqq t' < \eta d(z).$$

First choose a positive number δ' such that

$$t' < \eta(d(z') - \delta'). \qquad (8.13)$$

Second we choose a compact subset K of G such that the following two conditions are satisfied:

a) the point z' is an inner point of K,

b) the distance δ of K from the boundary ∂G of G is not larger than δ'.

Define $\tilde{K}$ by (8.11) with this choice of K and δ. The intersection of $\tilde{K}$ with the hyperplane t = t' is characterized by

$$\frac{t'}{\eta} + \delta \leqq d(z).$$

In view of (8.13) we have further

$$\frac{t'}{\eta} + \delta \leqq \frac{t'}{\eta} + \delta' < d(z').$$

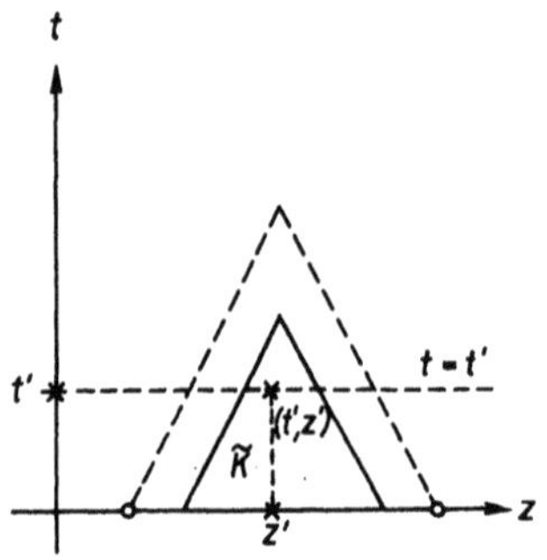

Thus the point (t',z') is an inner point of the intersection of $\tilde{K}$ with the hyperplane $t = t'$.

Now we are able to prove that $H_*(M)$ is complete. Take any fundamental sequence $\{w_n\}_{n=1,2,\ldots}$ in $H_*(M)$, i.e.,

$$\|w_n - w_m\|_* < \varepsilon$$

for n, m large enough. Thus everywhere in M the inequality

$$|w_n(t,z) - w_m(t,z)|d^p(t,z) < \varepsilon \tag{8.14}$$

holds. Any point (t',z') of M belongs to a compact subset $\tilde{K}$ defined by (8.11). Taking into account the estimate (8.12), the inequality (8.14) yields

$$|w_n(t,z) - w_m(t,z)| < \frac{\varepsilon}{\delta^p}$$

provided (t,z) belongs to $\tilde{K}$. Therefore, the given fundamental sequence converges uniformly in each compact subset $\tilde{K}$ of the type (8.11). This implies that the limit function $w = w(t,z)$ is continuous and depends holomorphically on z. Since each point of M belongs to a suitably chosen compact set $\tilde{K}$ the last statement is true for all points of M. Carrying out the limiting process $m \to \infty$ in (8.14), one obtains

$$|w_n(t,z) - w(t,z)|d^p(t,z) \leq \varepsilon \tag{8.15}$$

for each (t,z) belonging to M. Since

$$|w(t,z)|d^p(t,z)$$
$$\leq |w(t,z) - w_n(t,z)|d^p(t,z) + |w_n(t,z)|d^p(t,z) \leq \varepsilon + \|w_n\|_*$$

the limit function $w = w(t,z)$ itself belongs to $H_*(M)$. The estimate (8.15) implies

$$\|w_n - w\|_* \leq \varepsilon.$$

Thus the $w_n = w_n(t,z)$ do not only converge pointwise to $w = w(t,z)$ but also in the norm of $H_*(M)$. In this way we have proved the completeness of $H_*(M)$.

Now we intend to formulate Nagumo's lemma for functions belonging to $H_*(M)$. For the sake of brevity we restrict ourselves to the case that z is a point in the complex plane but not in $\mathbb{C}^n$.

Take any point (t,z) of M. Then we have

$$d(t,z) = d(z) - \frac{t}{\eta} > 0.$$

Choose any positive number r with $r < d(t,z)$ and regard all points z' with $|z' - z| = r$. In view of the triangle inequality we have

$$d(z') \geq d(z) - r$$

and consequently,

$$d(t,z') = d(z') - \frac{t}{\eta} \geq d(z) - \frac{t}{\eta} - r = d(t,z) - r > 0 \tag{8.16}$$

This estimate yields, first, that

$$\frac{t}{\eta} < d(z').$$

Hence the points (t,z') belong to M, too. From the definition of $\|\cdot\|_*$ we obtain immediately

$$|w(t,z')| \leq \frac{\|w\|_*}{d^p(t,z')} .$$

Once more taking into account the estimate (8.16), one gets

$$|w(t,z')| \leq \frac{\|w\|_*}{(d(t,z) - r)^p} .$$

Cauchy's integral formula for the first derivative leads, consequently, to the estimate

$$\left|\frac{\partial w}{\partial z}(t,z)\right| \leq \frac{1}{2\pi}\frac{1}{r^2}\frac{\|w\|_*}{(d(t,z) - r)^p} 2\pi r \leq \frac{1}{r}\frac{\|w\|_*}{(d(t,z) - r)^p} .$$

Choosing $r = \frac{1}{1 + p}d(t,z)$, the last inequality passes into the following one which is the desired Nagumo lemma for functions belonging to $H_*(M)$ (cf. [44]):

<u>Lemma 2.</u> If $w \in H_*(M)$, then

$$\left|\frac{\partial w}{\partial z}(t,z)\right| \leq \frac{C_p}{d^{p+1}(t,z)}\|w\|_* .$$

In the case of several complex variables $z_1, \ldots, z_n$ the last inequality is to be replaced by

$$\left|\frac{\partial w}{\partial z_j}(t,z)\right| \leq \frac{C_p}{d^{p+1}(t,z)}\|w\|_* .$$

Now we look for m complex-valued functions $w_j = w_j(t, z_1, \ldots, z_n)$, $j = 1, \ldots, m$, depending on the real variable t as well as n complex variables $z_1, \ldots, z_n$. The desired functions are to satisfy the system of partial differential equations

$$\frac{\partial w_j}{\partial t} = \sum_{i,k} a_{ik}^{(j)}(t,z)\frac{\partial w_k}{\partial z_i} + \sum_k a_k^{(j)}(t,z)w_k + a_0^{(j)}(t,z) \tag{8.17}$$

and the initial conditions

$$w_j(0,z) = \phi_j(z), \quad z = 1, \ldots, m, \tag{8.18}$$

as well, where the initial functions ϕ_j are holomorphic. The coefficients $a_{ik}^{(j)}$, $a_k^{(j)}$, and $a_0^{(j)}$ are supposed to be continuous in (t,z) and holomorphic in z. Assume that the coefficients are given in a conical domain M constructed above whose basis G is a bounded domain in the z-plane or in $\mathbb{C}^n$.

In order to solve the initial value problem (8.17), (8.18) we notice, first, that this problem is equivalent to the integro-differential equation

$$w_j(t,z) = \phi_j(z) + \int_0^t \Big[\sum_{i,k} a_{ik}^{(j)}(\tau,z) \frac{\partial w_k}{\partial z_i}(\tau,z) + \sum_k a_k^{(j)}(\tau,z) w_k(\tau,z) + a_0^{(j)}(\tau,z) \Big] d\tau. \tag{8.19}$$

We look for a solution of (8.19) belonging to the space $H_*(M)$, where the corresponding numbers p and η are fixed positive numbers (η will be chosen later).

Now define the following operator: Let $w = (w_1, \ldots, w_m)$ be any element of $H_*(M)$, $p > 0$. To this element w we assign the vector $W = (W_1, \ldots, W_m)$, where W_j is defined by the right-hand side of (8.19):

$$W_j(t,z) = \phi_j(z) + \int_0^t [\ldots] d\tau. \tag{8.20}$$

It is clear that a fixed point of the operator defined by (8.20) turns out to be a solution of the integro-differential equation (8.19) and, consequently, of the initial value problem (8.17), (8.18). The existence of a fixed point will be proved by using Banach's fixed-point theorem. For this end we have to estimate the operator defined by (8.20). We begin with suitable assumptions on the coefficients $a_{ik}^{(j)}$, $a_k^{(j)}$, $a_0^{(j)}$ of the differential equation (8.17). Suppose that there are constants A_*, B_*, C_*, and D_* such that

$$|a_{ik}^{(j)}(t,z)| \leq A_*, \quad |a_k^{(j)}(t,z)| \leq \frac{B_*}{d(t,z)}, \quad |a_0^{(j)}(t,z)| \leq \frac{C_*}{d^{p+1}(t,z)}, \quad |\phi_j(z)| \leq \frac{D_*}{d^p(z)} \tag{8.21}$$

for each point of M. Next we formulate some lemmas in order to comment upon these assumptions. In the following it is supposed that there exists a positive number d_0 such that $d(z) \leq d_0$ for each point of G. In this case the conical domain is bounded, too, and d(t,z) can be estimated by $d(t,z) \leq d_0$. Thus one gets $d^\alpha(t,z) \leq d_0^\alpha$, i.e.,

$$1 \leq \frac{d_0^\alpha}{d^\alpha(t,z)}$$

if α is any positive number. Consequently the following statement is true:

Lemma 3. Suppose that $f = f(t,z)$ is bounded in M,

$$|f(t,z)| \leq \text{const}.$$

Then f satisfies also the inequality

$$|f(t,z)| \leq \frac{\text{const}\cdot d_0^\alpha}{d^\alpha(t,z)} .$$

In view of this lemma the coefficients $a_{ik}^{(j)}$, $a_k^{(j)}$, $a_0^{(j)}$ satisfy a condition of type (8.21) provided they are bounded in M. Further the lemma shows that the third condition (8.21) is satisfied with $p > 0$ if only it is satisfied with $p = 0$. Furthermore, in view of $d(z) \leq d_0$ we have

$$1 \leq \frac{d_0^p}{d^p(z)} .$$

Therefore the fourth condition (8.21) is satisfied if ϕ_j is bounded.

Since

$$d(t,z) = d(z) - \frac{t}{\eta} \leq d(z)$$

we have

$$\frac{1}{d^\alpha(z)} \leq \frac{1}{d^\alpha(t,z)}$$

and, therefore, the following assertion is true:

Lemma 4. The fourth inequality (8.21) implies

$$|\phi_j(z)| \leq \frac{D_*}{d^p(t,z)} .$$

Once more taking into consideration the definition of $d(t,z)$, we obtain

$$\int_0^t \frac{d\tau}{d^{p+1}(\tau,z)} = \int_0^t \frac{d\tau}{(d(z) - \frac{\tau}{\eta})^{p+1}} = \frac{\eta}{p}\left[\frac{1}{(d(z) - \frac{\tau}{\eta})^p}\right]_0^t$$

$$= \frac{\eta}{p}\left[\frac{1}{(d(z) - \frac{t}{\eta})^p} - \frac{1}{d^p(z)}\right] < \frac{\eta}{p}\,\frac{1}{(d(z) - \frac{t}{\eta})^p} .$$

Thus also the following lemma holds:

Lemma 5. If $p > 0$, then

$$\int_0^t \frac{d\tau}{d^{p+1}(\tau,z)} < \frac{\eta}{p}\,\frac{1}{d^p(t,z)} .$$

Analogously we have

Lemma 6. If $0 < \alpha < 1$, then

$$\int_0^t \frac{d\tau}{d^\alpha(\tau,z)} < \frac{\eta}{1-\alpha} d_0^{1-\alpha} .$$

Such estimate is true because

$$\int_0^t \frac{d\tau}{(d(z) - \frac{\tau}{\eta})^\alpha} = - \frac{\eta}{1-\alpha}\left[(d(z) - \frac{\tau}{\eta})^{1-\alpha}\right]_0^t < \frac{\eta}{1-\alpha} d^{1-\alpha}(z).$$

It may be added that also the equation

$$\int_0^t \frac{d\tau}{d(\tau,z)} = \eta \log \frac{d(z)}{d(t,z)}$$

holds, but we shall not make use of it.

Now take any element $w = (w_1, \ldots, w_m)$ of $H_*(M)$. Hence the $w_k(z)$ can be estimated by

$$|w_k(t,z)| \leq \frac{\|w\|_*}{d^p(t,z)} .$$

Using Nagumo's lemma for $H_*(M)$, the last estimate yields

$$\left|\frac{\partial w_k}{\partial z_i}(t,z)\right| \leq \frac{C_p}{d^{p+1}(t,z)}\|w\|_* .$$

By virtue of the assumption (8.21) the modulus of the integrand in formula (8.20) is, consequently, not greater than

$$[(nmA_*C_p + mB_*)\|w\|_* + C_*]\frac{1}{d^{p+1}(\tau,z)} .$$

Applying lemma 5, the modulus permits an estimate of type

$$\frac{\text{const}}{d^p(t,z)} .$$

Lemma 3 implies an analogous estimate for the modulus of the first term on the right-hand side of (8.20). Therefore the image $W = (W_1, \ldots, W_m)$ of $w = (w_1, \ldots, w_m)$ belongs to $H_*(M)$, too, i.e., the following theorem holds:

Theorem 2. The operator defined by (8.20) maps $H_*(M)$ into itself.

Now we look for conditions guaranteeing that the operator under consideration turns out to be a contraction. For this end we start from two elements w and $\tilde{w}$ belonging to $H_*(M)$. Replacing w on the right-hand side of (8.20) by $\tilde{w}$, we get the definition of $\tilde{W}_j(t,z)$. First we are going to estimate the norm of $W - \tilde{W}$, where $\tilde{W} = (\tilde{W}_1, \ldots, \tilde{W}_m)$. Subtracting this formula from the corresponding formula (8.20), one obtains a representation of $W_j - \tilde{W}_j$ by an integral. Once more applying Nagumo's lemma, it follows that the modulus of the integrand can be estimated by

$$(nmA_*C_p + mB_*)\frac{\|w - \tilde{w}\|_*}{d^{p+1}(t,z)}$$

(note that both ϕ_j and $A_0^{(j)}$ neutralize each other in the difference). Applying lemma 4 and taking into consideration the definition of the norm in $H_*(M)$, the last estimate yields

$$\|W - \tilde{W}\|_* \leq m(nA_*(1 + \frac{1}{p})^{p+1} + \frac{B_*}{p})\eta\|w - \tilde{w}\|_* .$$

Therefore the operator defined by (8.20) turns out to be contractive provided η satisfies the inequality

$$m(\eta A_*(1 + \frac{1}{p})^{p+1} + \frac{B_*}{p}) < \frac{1}{\eta} . \qquad (8.22)$$

Applying Banach's fixed-point theorem, the following statement has been obtained:

Theorem 3. If η satisfies condition (8.22), then there exists a solution $w = w(t,z)$ of the initial value problem (8.17), (8.18) defined globally in the whole domain M.

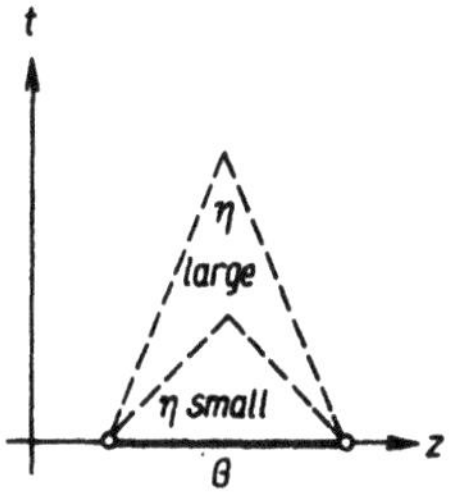

Concluding our considerations on the linear Cauchy-Kovalevskaya system (8.17), we would like to remark that condition (8.22) can always be satisfied by choosing η small enough. Notice further that a diminuation of η results in a diminuation of the conical domain M, i.e., its height decreases. Further note that the solution is uniquely determined at least in $H_*(M)$ because Banach's fixed-point theorem implies the uniqueness of fixed points, too.

Now we are going to solve the general nonlinear initial value problem (4.4), (4.5) by using W. Walter's method of the modified supremum norm (8.10). As it was done in section 4.1. we suppose that the right-hand side $f(t,z,w,p)$ of (4.4) is continuous in all its variables and depends holomorphically on z, and the components of the vectors w and p as well. Analogously the initial vectors $\phi^{(o)}, \ldots, \phi^{(k-1)}$ in (4.5) are supposed to be holomorphic in z. By virtue of section 4.2. we reduce the given initial value problem (4.4), (4.5) to the following one:

We look for a solution $w = (w_1, \ldots, w_m)$ of the quasilinear system (4.17) taking identically vanishing initial values at $t = 0$:

$$w_j(0,z) = 0 \quad \text{for each } z.$$

The coefficients $a_{ik}^{(j)}$, $a_o^{(j)}$ of (4.17) depend on t, z, and w. Let G be a given bounded domain in the z-space and M the above-constructed conical domain (with the basis G) depending on the choice of η. Assume that the coefficients are defined in the set

$$\{(t,z,w) : (t,z) \in M, |w| \leq R\}, \qquad (8.23)$$

where R is a given positive number and $|w|$ means $\max_j |w_j|$. Suppose further that the coefficients $a_{ik}^{(j)}$, $a_o^{(j)}$ are continuous in (t,z,w) and holomorphic in z and w. Finally assume that the given coefficients satisfy conditions of the type

$$|a_{ik}^{(j)}(t,z,w)| \leq A_*, \qquad (8.24)$$

$$|a_{ik}^{(j)}(t,z,w) - a_{ik}^{(j)}(t,z,\tilde{w})| \leq \frac{L_*|w - \tilde{w}|}{\sqrt{d(t,z)}}, \tag{8.25}$$

$$|a_o^{(j)}(t,z,w)| \leq \frac{C_*}{\sqrt{d(t,z)}}, \tag{8.26}$$

$$|a_o^{(j)}(t,z,w) - a_o^{(j)}(t,z,\tilde{w})| = \frac{L_{**}|w - \tilde{w}|}{d(t,z)} \tag{8.27}$$

in the set (8.23), where A_*, C_*, L_*, L_{**} are given constants.

Note that in view of lemma 3 the condition (8.26) is satisfied if $a_o^{(j)}$ is bounded. Analogously, in view of the same lemma 3 the modified Lipschitz conditions (8.25) and (8.27) are satisfied provided the coefficients $a_{ik}^{(j)}$ and $a_o^{(j)}$ are Lipschitz continuous in the ordinary sense.

Next notice that the initial value problem under consideration, i.e., the system (4.17) with homogeneous initial data, is equivalent to the system of integro-differential equations

$$w_j(t,z) = \int_0^t \Big[\sum_{i,k} a_{ik}^{(j)}(\tau,z,w(\tau,z))\frac{\partial w_k}{\partial z_i}(\tau,z) + a_o^{(j)}(\tau,z,w(\tau,z))\Big]d\tau, \tag{8.28}$$

$j = 1, \ldots, m$. In order to solve this system, we look for a fixed point of the operator defined by the right-hand side of (8.28), i.e.,

$$W_j(t,z) = \int_0^t [\ldots]d\tau. \tag{8.29}$$

The corresponding operator (8.20) in the linear case (8.17) is defined in the whole space $H_*(M)$. Thus in the linear case we were able to apply the operator (8.20) to the whole space $H_*(G)$. On the contrary, the operator (8.29) is not defined in the whole space $H_*(M)$ because the coefficients $a_{ik}^{(j)}$, $a_o^{(j)}$ are defined only if $|w| \leq R$ (cf. (8.23)). In order to apply Banach's fixed-point theorem we must, consequently, find a closed subset of $H_*(M)$ in which the operator (8.29) is defined everywhere. For this end we define the subset S of $H_*(M)$ by

$$S = \left\{ w \in H_*(M) \; : \; |w| \leq R, \quad \left|\frac{\partial w_k}{\partial z_i}\right| \leq \frac{1}{\sqrt{d(t,z)}}\right\}.$$

In order to show that S is a closed subset of $H_*(M)$, we take any sequence $\{w^{(\nu)}\}_{\nu=1,2,\ldots}$ in S which converges with respect to the metric of $H_*(M)$. We have to prove that its limit element w^* belongs to S, too. Since all the $w^{(\nu)} = (w_1^{(\nu)}, \ldots, w_m^{(\nu)})$ belong to S we have

$$\left|w_k^{(\nu)}\right| \leq R, \qquad \left|\frac{\partial w_k^{(\nu)}}{\partial z_i}\right| \leq \frac{1}{\sqrt{d(t,z)}} \tag{8.30}$$

for each $\nu = 1, 2, \ldots$ On the other hand, we know that convergence in $H_*(M)$ implies pointwise convergence. Thus the inequalities (8.30) are also satisfied for the limit w^*. This means that w^* belongs to S and S is proved to be closed.

It is clear that the operator (8.29) is defined for each w belonging to S. First we look for conditions under which the integro-differential operator (8.29) maps the set S into itself. For this end we estimate the images W_j of S as well as their derivatives $\frac{\partial W_j}{\partial z_1}$ for an arbitrary element w of S. In view of (8.24) and (8.26) the module of the integrand in (8.29) does not exceed

$$nmA_* \frac{1}{\sqrt{d(t,z)}} + \frac{C_*}{\sqrt{d(t,z)}} .$$

Taking into account lemma 6, we get, therefore, the estimate

$$|W_j| \leq 2\eta(nmA_* + C_*)d_0^{1/2}.$$

Hence the images W_j may be estimated by $|W_j| \leq R$ provided η satisfies the inequality

$$2\eta(nmA_* + C_*)d_0^{1/2} \leq R. \tag{8.31}$$

It remains to estimate the derivatives $\frac{\partial W_j}{\partial z_1}$. Regard the composed functions defined by

$$a_{ik}^{(j)}(t,z,w(t,z)), \quad a_o^{(j)}(t,z,w(t,z))$$

and denote their derivatives with respect to z_1 by

$$\frac{\partial}{\partial z_1}a_{ik}^{(j)}(\ldots) \quad \text{and} \quad \frac{\partial}{\partial z_1}a_o^{(j)}(\ldots) \quad \text{resp.}$$

Then formula (8.29) implies immediately

$$\frac{\partial W_j}{\partial z_1}(t,z) = \tag{8.32}$$
$$\int_0^t \Big[\sum_{i,k}\Big(\frac{\partial}{\partial z_1}a_{ik}^{(j)} \ (\ldots) \ \frac{\partial w_k}{\partial z_i} + a_{ik}^{(j)} \ (\ldots) \ \frac{\partial^2 w_k}{\partial z_i \partial z_1}\Big) + \frac{\partial}{\partial z_1}a_o^{(j)} \ (\ldots)\Big]d\tau,$$

where the dots are to replace by τ, z, $w(\tau,z)$. Since $w = w(t,z)$ belongs to S we have

$$\left|\frac{\partial w_k}{\partial z_i}(t,z)\right| \leq \frac{1}{\sqrt{d(t,z)}} .$$

Applying Nagumo's lemma with $p = 0.5$, one obtains

$$\left|\frac{\partial^2 w}{\partial z_i \partial z_1}(t,z)\right| \leq \frac{C_{0.5}}{d^{1.5}(t,z)} .$$

Once more applying Nagumo's lemma, from the assumptions (8.24) and (8.26) one gets, analogously,

$$\left|\frac{\partial}{\partial z_1}a_{ik}^{(j)} \ (\ldots)\right| \leq \frac{A_*}{d(t,z)} ,$$

$$\left|\frac{\partial}{z_1} a_o^{(j)} \; (\ldots)\right| \leq \frac{C_* C_{0.5}}{d^{1.5}(t,z)} .$$

The module of the integrand in (8.32) is, therefore, not larger than

$$(nmA_*(1 + C_{0.5}) + C_* C_{0.5})\frac{1}{d^{1.5}(t,z)} ,$$

and lemma 5 yields

$$\left|\frac{\partial W_j}{\partial z_1}\right| \leq 2\eta(nmA_*(1 + C_{0.5}) + C_* C_{0.5})\frac{1}{\sqrt{d(t,z)}} .$$

Thus the images W_j satisfy the inequality

$$\left|\frac{\partial W_j}{\partial z_1}\right| \leq \frac{1}{\sqrt{d(t,z)}}$$

provided we assume that

$$2\eta(nmA_*(1 + C_{0.5}) + C_* C_{0.5}) \leq 1. \tag{8.33}$$

Summarizing the above considerations, the following lemma has been proved:

<u>Lemma 7.</u> Suppose that η satisfies the inequalities (8.31) and (8.33). Then the integro-differential operator (8.29) maps S into itself.

Second we ask under which condition the operator (8.29) is contractive. To it we choose arbitrarily two elements $w = w(t,z)$ and $\tilde{w} = \tilde{w}(t,z)$ belonging to the closed subset S of $H_*(M)$, $p > 0$. Analogously to (8.29) the image $\tilde{W} = (\tilde{W}_1, \ldots, \tilde{W}_m)$ of $\tilde{w}$ is defined by

$$\tilde{W}_j(t,z) = \int_0^t \left[\sum_{i,k} a_{ik}^{(j)} \; (\ldots) \; \frac{\partial \tilde{w}_k}{\partial z_i} \; (\ldots) + a_o^{(j)} \; (\ldots)\right] d\tau,$$

where the variables of the $a_{ik}^{(j)}$, $a_o^{(j)}$ are $(\tau, z, \tilde{w}(\tau,z))$. Now regard $W_j(t,z) - \tilde{W}_j(t,z)$. The integrand of the integral representation of this difference may be written in the form

$$\begin{aligned} &\sum_{i,k}\left\{a_{ik}^{(j)}(\tau,z,w(\tau,z)) - a_{ik}^{(j)}(\tau,z,\tilde{w}(\tau,z))\right\}\frac{\partial w_k}{\partial z_i}(\tau,z) \\ &+ \sum_{i,k} a_{ik}^{(j)}(\tau,z,\tilde{w}(\tau,z))\left(\frac{\partial w_k}{\partial z_i}(\tau,z) - \frac{\partial \tilde{w}_k}{\partial z_i}(\tau,z)\right) \\ &+ \left\{a_o^{(j)}(\tau,z,w(\tau,z)) - a_o^{(j)}(\tau,z,\tilde{w}(\tau,z))\right\}. \end{aligned} \tag{8.34}$$

The definition (8.10) of the norm $\|\cdot\|_*$ implies immediately

$$|w(\tau,z) - \tilde{w}(\tau,z)| \leq \frac{\|w - \tilde{w}\|_*}{d^p(\tau,z)} . \tag{8.35}$$

Since w belongs to S one has further

$$\left|\frac{\partial w_k}{\partial z_i}(\tau,z)\right| \leq \frac{1}{\sqrt{d(\tau,z)}} .$$

In view of (8.25) the module of the first term in (8.34) can be estimated by

$$nmL_* \frac{1}{d^{p+1}(\tau,z)} \|w - \tilde{w}\|_* . \tag{8.36}$$

Analogously, the assumption (8.27) implies that the modulus of the third term of (8.34) is not larger than

$$L_{**} \frac{1}{d^{p+1}(\tau,z)} \|w - \tilde{w}\|_* . \tag{8.37}$$

Using Nagumo's lemma, the estimate (8.35) yields

$$\left| \frac{\partial w_k}{\partial z_i}(\tau,z) - \frac{\partial \tilde{w}_k}{\partial z_i}(\tau,z) \right| \leq \frac{C_p}{d^{p+1}(\tau,z)} \|w - \tilde{w}\|_* .$$

Taking into account the assumption (8.24), the module of the second term in (8.34) may be estimated by

$$nmA_* C_p \frac{1}{d^{p+1}(\tau,z)} \|w - \tilde{w}\|_* . \tag{8.38}$$

The sum of (8.36), (8.37), and (8.38) is a bound of the integrand in the integral representation formula for $W_j(t,z) - \tilde{W}_j(t,z)$. Applying lemma 5, it follows, therefore, that

$$\begin{aligned} &|W_j(t,z) - \tilde{W}_j(t,z)| \\ &\leq \left\{ nm(A_* C_p + L_*) + L_{**} \right\} \frac{\eta}{p} \frac{1}{d^p(t,z)} \|w - \tilde{w}\|_* . \end{aligned}$$

Once more taking into consideration the definition (8.10), the last estimate may be rewritten in the form

$$\|W - \tilde{W}\|_* \leq \eta \left\{ nm(1 + \tfrac{1}{p})^{p+1} A_* + \tfrac{1}{p}(nmL_* + L_{**}) \right\} \|w - \tilde{w}\|_*$$

since $C_p = (p + 1)(1 + \frac{1}{p})^p$. Thus the operator (8.29) is contractive if

$$\eta \left\{ nm(1 + \tfrac{1}{p})^{p+1} A_* + \tfrac{1}{p}(nmL_* + L_{**}) \right\} < 1 . \tag{8.39}$$

According to Banach's fixed-point theorem, the following theorem is true:

Theorem 4. Suppose that the coefficients $a_{ik}^{(j)}$, $a_o^{(j)}$ of the quasilinear system (4.17) satisfy the conditions (8.24), (8.25), (8.26), (8.27), where A_*, C_*, L_*, L_{**} are given constants. Assume further that $p > 0$ is fixedly chosen. Choose η so small that the inequalities (8.31), (8.33), and (8.39) are satisfied[1]). Then the initial value problem

$$w_j(0,z) = 0$$

for the system (4.17) possesses a uniquely determined solution in the subset S of $H_*(M)$.

---

[1]) This η must not exceed, of course, the originally given one, characterizing the conical set M in which the coefficients of (4.17) are defined.

8.2. Generalized analytic functions depending on time in conical domains

Modifying W. Walter's method (see section 8.1.), in the present section we are going to construct generalized analytic functions depending on time and satisfying a given initial condition (cf. [74]). For this end it is not possible to make use of the weighted maximum norm (8.10) by reason of the unboundedness of the Π_G-operator in the space $C(\overline{G})$ equipped with the maximum norm.

For the sake of simplicity assume that G is a bounded domain in the z-plane. We first choose, in a similar way as done in section 7.4., a family of subdomains G_s of G, where the index s runs in an open interval, $0 < s < s_0$. In contradistinction from the properties a), b), and c) of a family of subdomains formulated in section 2.2., for the purpose of the present section we need a family of subdomains satisfying the following conditions:

a) With the exception of one point z_0 of G to each point z of G there exists a uniquely determined s(z) with $0 < s(z) < s_0$ such that $z \in \partial G_{s(z)}$, i.e., z belongs to the boundary of exactly one subdomain $G_{s(z)}$ of the given family.

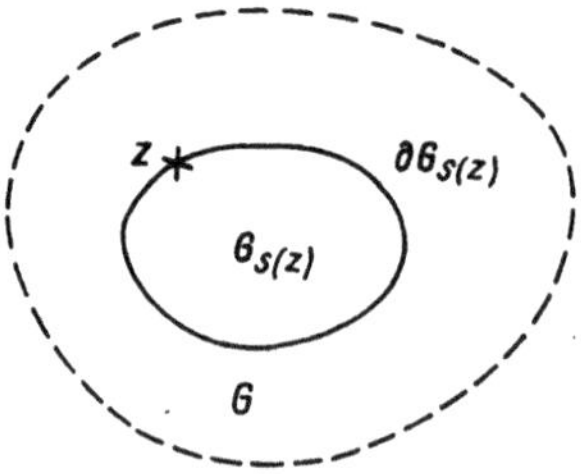

b) The closure $\overline{G}_{s'}$ is a compact subset of G_s if only $s' < s$.

c) The distance of $G_{s'}$ from the boundary ∂G_s of G_s can be estimated by

$$\text{dist}\,(G_{s'}, \partial G_s) \geq \text{const}(s - s'),$$

where the constant is independent of s' and s, $s' < s$.

d) For all s, $0 < s < s_0$, the norms $\|T_{G_s}\|$ and $\|\Pi_{G_s}\|$ of the T_{G_s}- and Π_{G_s}-operators interpreted as operators in the space of Hölder-continuous functions are bounded, i.e., there exist constants C_1 and C_2 not depending on s such that

$$\|T_{G_s}\| \leq C_1, \quad \|\Pi_{G_s}\| \leq C_2.$$

As an example for a family of subdomains satisfying the above-formulated four conditions regard the case of a disk with radius R centred at z_0. Then the disks with radii sR, $0 < s < 1$, form a family of subdomains of the described kind. The proof of theorem 2 of section 12.1. of the book [64] shows that the norms $\|T_{G_s}\|$ are bounded, whereas the boundedness of the norms $\|\Pi_{G_s}\|$ follows immediately from lemma 7 of section 13.3. of the same book (cf. also the corresponding annotation in section 7.4. of the present book, p. 123).

Notice that the number $s(z)$ existing in accordance with condition a) may be used in order to characterize the distance of z from the boundary of the given domain G: the distance may be defined as difference $s_0 - s(z)$. Then the boundaries $\partial G_{s(z)}$ are level curves of that distance. The exceptional point z_0 has the distance s_0 from ∂G.

Using this kind of distance, we define the following conical domain in the t,z-space

$$M = \{(t,z) : z \in G, \quad 0 \leqq t < \eta(s_0 - s(z))\}$$

which replaces the corresponding conical domain M introduced in the preceding section 8.1. The number η will be fixed later. The non-negative value of the expression

$$d(t,z) := s_0 - s(z) - \frac{t}{\eta}$$

measures the distance of (t,z) from the lateral surface of M. The intersection of M with the hyperplane $t = t'$ is

$$\{(t',z) : z \in G_{s'}, \; s' := s_0 - \frac{t'}{\eta}\}.$$

Take any δ, $0 < \delta < s_0$, and define $\tilde{K}$ by

$$\tilde{K} = \{(t,z) : z \in G, \quad 0 \leqq t \leqq \eta(s_0 - s(z) - \delta)\}.$$

Then $\tilde{K}$ is a compact subset of M whose points satisfy the inequality

$$d(t,z) \geqq \delta. \qquad (8.40)$$

The intersection of $\tilde{K}$ with the hyperplane $t = t'$ contains all points (t',z) with $z \in \overline{G}_{s'}$, where

$$s' = s_0 - \delta - \frac{t'}{\eta}.$$

Now let l be a differential operator of type (6.59), and regard the differential equation

$$lw = \frac{\partial w}{\partial \overline{z}} - a(z)w - b(z)\overline{w} = 0. \qquad (8.41)$$

Suppose the coefficients $a(z)$, $b(z)$ are defined and Hölder-continuous in $\overline{G}$ with Hölder-exponent (index) α, $0 < \alpha < 1$. Again denote the Hölder-norm in $C^\alpha(\overline{G}_s)$ by $\|\cdot\|_s$ (cf. section 7.4.). Then we have

$$\|a(z)\|_s \leqq \|a(z)\|_{C^\alpha(\overline{G})}, \quad \|b(z)\|_s \leqq \|b(z)\|_{C^\alpha(\overline{G})}.$$

Note that the following considerations can also be carried out if only $a(z)$, $b(z)$ are Hölder-continuous in G, not in $\overline{G}$. Then we have, additionally, to demand that the norms $\|a(z)\|_s$ and $\|b(z)\|_s$ are bounded, i.e., that there exist constants a_* and b_* such that

$$\|a(z)\|_s \leqq a_*, \quad \|b(z)\|_s \leqq b_*$$

for every s, $0 < s < s_0$. In this case in the following considerations the norms $\|a(z)\|_{C^\alpha(\overline{G})}$ and $\|b(z)\|_{C^\alpha(\overline{G})}$ resp. are to be replaced by a_* and b_* resp. (see [74]).

Regard continuous functions $w = w(t,z)$ defined in M and Hölder-continuous in $\overline{G}_s$ for fixed $t = t'$ (with a fixed Hölder index α smaller than 1) provided $s < s'$, where s' is defined by

$$s' = s_0 - \frac{t'}{\eta} .$$

From the definition of M we obtain that

$$s(z) < s_0 - \frac{t'}{\eta} ,$$

i.e., $w = w(t',z)$ is Hölder-continuous in $\overline{G}_{s(z)}$, especially. Therefore,

$$\|w(t,z)\|_{s(z)}$$

is defined for each z ($\neq z_0$).

Let $p > 0$ be a fixed number, and regard all those functions $w = w(t,z)$ for which

$$\|w(t,z)\|_{s(z)} d^p(t,z)$$

is bounded. Denote

$$\sup_M \|w(t,z)\|_{s(z)} d^p(t,z) \tag{8.42}$$

by $\|w\|_*$, where the points (z_0,t), $0 \leq t < s_0$, are to omit in the last supremum.

Analogously to $H_*(M)$ (see 8.1.) define $W_*(M)$ as space of all functions $w = w(t,z)$ with finite norm $\|w\|_*$ for which $w = w(t,z)$ is solution of the differential equation (8.41) for fixed t.

Notice that the definition (8.42) is a generalization of (8.10) to the case of Hölder-continuous functions. In a similar way as (8.10) defines a norm in $H_*(M)$, the functional (8.42) defines a norm in $W_*(M)$.

The definition (8.42) of the norm $\|w\|_*$ and the estimate (8.40) lead to the inequality

$$\|w(t,z)\|_{s(z)} \leq \frac{\|w\|_*}{\delta^p}$$

for points belonging to $\tilde{K}$. Applying this inequality to a fundamental sequence in $W_*(M)$, i.e., to a sequence $\{w_n\}_{n=1,2,...}$ with

$$\|w_n - w_m\|_* < \varepsilon$$

for sufficiently large n and m, one obtains

$$\|w_n(t,z) - w_m(t,z)\|_{s(z)} < \frac{\varepsilon}{\delta^p} \tag{8.43}$$

for those n, m. Taking into account the definition 6.2.6. of the Hölder-norm, we see

$$|w_n(t,z) - w_m(t,z)| \leq \|w_n(t,z) - w_m(t,z)\|_{s(z)}$$

and, consequently,

$$|w_n(t,z) - w_m(t,z)| < \frac{\varepsilon}{\delta^p}$$

for sufficiently large n, m and $(t,z) \in \tilde{K}$. This implies that a fundamental sequence in $W_*(M)$ converges uniformly in each compact subset $\tilde{K}$. Therefore, the limit function $w = w(t,z)$ is continuous in $\tilde{K}$.

Let (t,z) be an arbitrary point of M, i.e.,

$$0 \leq t < \eta(s_0 - s(z)).$$

Choose $\delta > 0$ such that

$$0 \leq t < \eta(s_0 - s(z) - \delta). \tag{8.44}$$

Take into consideration that s(z') belongs to a neighbourhood of s(z) if z' belongs to a neighbourhood of z. Therefore, the inequality (8.44) is also satisfied for (t',z'), i.e.,

$$0 \leq t' < \eta(s_0 - s(z') - \delta),$$

provided (t',z') belongs to a sufficiently small neighbourhood of (t,z) (if $t = 0$, then we restrict t' to a neighbourhood on the right of $t = 0$). Thus to each point (t,z) of M there exists a compact set $\tilde{K}$ containing a neighbourhood of this point, too (in the case $t = 0$ this statement is true only for those points (t',z') of a neighbourhood for which $t' \geq 0$).

The last consideration shows that the limit function is continuous everywhere in M.

Finally take a fixed $t' \geq 0$, $t' < \eta s_0$. Again define s' by

$$s' = s_0 - \frac{t'}{\eta}.$$

Take any $s < s'$. Then there exist $\delta > 0$ such that

$$s \leq s_0 - \frac{t'}{\eta} - \delta,$$

i.e.,

$$t' \leq \eta(s_0 - s - \delta)$$

and, consequently,

$$d(t',z) = s_0 - s - \frac{t}{\eta} \geq \delta$$

provided $s(z) \leq s$, i.e., $z \in \overline{G_s}$. Once more taking into account (8.43) (with $t' = t$), we see that the $w_n(t',z)$ form a fundamental sequence in $C^\alpha(\overline{G_s})$. The completeness of that space implies that the limit function belongs to the same space. Since all $w_n(t',z)$ are solutions of the differential equation, the same is true for their limit functions. Hence it follows that $W_*(M)$ is complete.

Let us recall that in a compact subset the Hölder-norm of the derivative Φ' of a holomorphic function Φ can be estimated by

$$\|\Phi'\|_K \leq \frac{C_3}{\delta}\|\Phi\|_{G'} \tag{8.45}$$

provided the distance of K from the boundary of G' is not smaller than δ (cf. 6.2.11., where the constant C turns out to be equal to $3\cdot 2^\alpha$). Using this estimate, we are now in a position to prove the following statement:

<u>Lemma.</u> If $w \in W_*(M)$, then the estimate

$$\left\|\frac{\partial w}{\partial \bar z}(t,z)\right\|_{s(z)} d^{p+1}(t,z) \leq c_p\|w\|_*$$

holds at each point (t,z) of M, where the constant c_p depends only on p, c_0, C_1, C_2, C_3, a_*, and b_*.

<u>Proof.</u> Let (t,z) be an arbitrary point of M, i.e.,

$$d(t,z) > 0.$$

Define

$$r := \frac{1}{p+1}d(t,z) \tag{8.46}$$

and

$$\tilde s := s(z) + r.$$

From the definition of $d(t,z)$ one gets immediately that

$$d(t,z) \leq s_0 - s(z)$$

and, consequently,

$$\tilde s \leq s(z) + \frac{1}{p+1}(s_0 - s(z)) = \frac{p}{p+1}s(z) + \frac{1}{p+1}s_0 < s_0$$

because $s(z) < s_0$. Choose any point $\tilde z$ on $\partial G_{\tilde s}$, i.e., $s(\tilde z) = \tilde s$. Then we have

$$d(t,\tilde z) = s_0 - \tilde s - \frac{t}{\eta} = s_0 - s(z) - \frac{t}{\eta} - r = d(t,z) - r$$

and, consequently,

$$\|w(t,z)\|_{\tilde s} \leq \frac{\|w\|_*}{d^p(t,\tilde z)} = \frac{|w|_*}{(d(t,z) - r)^p}\,. \tag{8.47}$$

Now define the function Φ depending on t and z by

$$\Phi = w - T_{G_{\tilde s}}(aw + b\bar w). \tag{8.48}$$

For fixed t one gets

$$\frac{\partial\Phi}{\partial\bar z} = \frac{\partial w}{\partial\bar z} - (aw + b\bar w) = 0,$$

i.e., Φ turns out to be holomorphic in $G_{\tilde s}$. Since

$$\|aw\|_{\tilde{s}} \le 2\|a\|\cdot\|w\|_{\tilde{s}} \le 2a_*\|w\|_{\tilde{s}},$$

$$\|bw\|_{\tilde{s}} \le 2\|b\|\cdot\|w\|_{\tilde{s}} \le 2b_*\|w\|_{\tilde{s}}$$

(cf. 6.3.9.) the estimate (8.47) yields

$$\begin{aligned}\|\Phi\|_{\tilde{s}} &\le \|w\|_{\tilde{s}} + \|T_{G_{\tilde{s}}}\|\cdot 2(a_* + b_*)\|w\|_{\tilde{s}} \\ &\le (1 + 2C_1(a_* + b_*))\frac{\|w\|_*}{(d(t,z) - r)^p} .\end{aligned} \tag{8.49}$$

Taking into account that the distance of $G_{s(z)}$ from the boundary of $G_{\tilde{s}}$ is at least

$$c_0(\tilde{s} - s(z)) = \frac{c_0}{p+1}d(t,z)$$

and using the a-priori estimate (8.45), one obtains for the function Φ defined by (8.48)

$$\|\Phi'\|_{s(z)} \le \frac{C_3(p+1)}{c_0 d(t,z)}\|\Phi\|_{\tilde{s}}. \tag{8.50}$$

The definition of Φ implies immediately

$$\frac{\partial w}{\partial \bar{z}} = \Phi' + \Pi_{G_{\tilde{s}}}(aw + b\bar{w}),$$

i.e.,

$$\left\|\frac{\partial w}{\partial \bar{z}}\right\|_{s(z)} \le \|\Phi'\|_{s(z)} + C_2\cdot 2(a_* + b_*)\|w\|_{\tilde{s}} \tag{8.51}$$

because

$$\|w\|_{s(z)} \le \|w\|_{\tilde{s}}.$$

Taking into consideration the definition (8.46) of r, one gets

$$\frac{1}{(d(t,z) - r)^p} = (1 + \frac{1}{p})^p\frac{1}{d^p(t,z)} .$$

Therefore, (8.47) passes into

$$\|w\|_{\tilde{s}} \le (1 + \frac{1}{p})^p\frac{\|w\|_*}{d^p(t,z)} , \tag{8.52}$$

while (8.50) and (8.49) lead to

$$\|\Phi'\|_{s(z)} \le \frac{C_3(p+1)}{c_0}(1 + 2C_1(a_* + b_*))(1 + \frac{1}{p})^p\frac{\|w\|_*}{d^{p+1}(t,z)} . \tag{8.53}$$

Since $d(t,z) \le s_0$ we have

$$\frac{1}{d^p(t,z)} \le \frac{s_0}{d^{p+1}(t,z)} .$$

Substituting (8.52) and (8.53) into (8.51), we obtain, consequently, the lemma, where the constant c_p is equal to

$$\left[\frac{p+1}{c_0}C_3(1 + 2C_1(a_* + b_*)) + 2s_0C_2(a_* + b_*)\right](1 + \frac{1}{p})^p.$$

Remark. Choosing

$$a(z) = 0, \quad b(z) = 0$$

everywhere in $\overline{G}$, one obtains the holomorphic case as special case of the differential equation (8.41). Then we have

$$a_* = b_* = 0,$$

c_p takes the value

$$\frac{p+1}{c_0} C_3 (1 + \frac{1}{p})^p,$$

and our lemma is a variant of the Nagumo lemma allowing us to estimate the Hölder-norm of the derivative of a holomorphic function, whereas the original Nagumo lemma (see 8.1., lemma 2) is an estimate of the supremum norm.

Using the above Nagumo lemma for the Hölder-norm of generalized analytic functions, we are going to construct the solution of initial value problems with generalized analytic initial functions by a contraction-mapping principle in $W_*(M)$. Regard again the initial value problem

$$\frac{\partial w}{\partial t} = C(t,z)\frac{\partial w}{\partial z} + A(t,z)w + B(t,z)\overline{w} + D(t,z), \tag{8.54}$$

$$w(0,z) = w_0(z) \tag{8.55}$$

which has already been investigated by the method of scales of Banach spaces in section 7.4. (there the initial value problem (8.54), (8.55) is marked with (7.37)). Again denote the differential operator on the right-hand side of (8.54) by $\tilde{L}$. Then the differential equation (8.54) can be rewritten as

$$\frac{\partial w}{\partial t} = \tilde{L}w$$

and the initial value problem (8.54), (8.55) is equivalent to the integro-differential equation

$$w(t,z) = w_0(z) + \int_0^t \tilde{L}w(\tau,z)d\tau \tag{8.56}$$

Consequently, each solution of the initial value problem (8.54), (8.55) turns out to be a fixed point of the operator defined by the right-hand side of (8.56), i.e., by

$$W(t,z) = w_0(z) + \int_0^t \tilde{L}w(\tau,z)d\tau. \tag{8.57}$$

Suppose the differential operator $\tilde{L}$ satisfies the following assumptions:

a) For every t the differential operator $\tilde{L}$ is associated to 1, i.e., $lw = 0$ implies $l(\tilde{L}w) = 0$ (cf. 6.4.1.).

b) The coefficients of $\tilde{L}$ are defined and continuous in the conical domain M. For fixed t they are Hölder-continuous in $\overline{G}_s$ provided $s < s_0 - \frac{t}{\eta}$. Suppose, moreover, that their Hölder-norms can be estimated by

$$\|C(t,z)\|_{s(z)} \leq C_*, \tag{8.58}$$

$$\|A(t,z)\|_{s(z)} \leq \frac{A_*}{d(t,z)}, \tag{8.59}$$

$$\|B(t,z)\|_{s(z)} \leq \frac{B_*}{d(t,z)}, \tag{8.60}$$

$$\|D(t,z)\|_{s(z)} \leq \frac{D_*}{d^{p+1}(t,z)}, \tag{8.61}$$

where A_*, B_*, C_*, and D_* are given constants.

c) The initial function $w_0 = w_0(z)$ is a Hölder-continuous solution to the associated differential equation (8.41), i.e., $lw_0 = 0$, and its Hölder-norm in $\overline{G}_s$, $s < s < s_0$, can be estimated by

$$\|w_0\|_{s(z)} \leq \frac{E_*}{(s_0 - s(z))^p}, \tag{8.62}$$

where E_* is a given constant, too.

Notice that $d(t,z) \leq s_0$ implies

$$1 \leq \frac{s_0^\alpha}{d^\alpha(t,z)}$$

for each positive exponent α. Therefore, the conditions (8.59), (8.60), and (8.61) are satisfied if the norms of the coefficients $A(t,z)$, $B(t,z)$, and $D(t,z)$ are uniformly bounded. Since

$$1 \leq \frac{s_0}{s_0 - s(z)}$$

one sees, analogously, that the condition (8.62) is satisfied if the norms $\|w_0\|_s$ are bounded for all s, $0 < s < s_0$.

Further, the condition (8.62) implies the estimate

$$\|w_0\|_{s(z)} \leq \frac{E_*}{d^p(t,z)} \tag{8.63}$$

because $s_0 - s(z) \geq d(t,z)$.

Note, finally, that $w = w(t,z) = 0$ everywhere in M is a special solution of the differential equation $lw = 0$ (8.41) for which we have

$$\tilde{L}w = D(t,z).$$

Therefore, by virtue of assumption a) the coefficient $D(t,z)$ turns out to be a solution of the associated differential equation $lw = 0$ (8.41).

Now take any $w = w(t,z)$ belonging to $W_*(M)$, i.e.,

$$\|w(\tau,z)\|_{s(z)} \leq \frac{\|w\|_*}{d^p(t,z)} . \tag{8.64}$$

Then the Nagumo lemma for the Hölder-norm of generalized analytic functions implies that

$$\left\|\frac{\partial w}{\partial z}(t,z)\right\|_{s(z)} \leq \frac{c_p}{d^{p+1}(t,z)}\|w\|_* . \tag{8.65}$$

Taking into account the assumptions (8.58) - (8.61), the estimates (8.64) and (8.65) imply

$$\|\tilde{L}w(t,z)\|_{s(z)} \leq (2(C_* c_p + A_* + B_*)\|w\|_* + D_*)\frac{1}{d^{p+1}(t,z)} .$$

Applying lemma 5 of 8.1. and taking into account the assumption (8.63), we get, consequently, for the image W(t,z) of w(t,z) defined by (8.57) the estimate

$$\|W(t,z)\|_{s(z)} \leq \left(E_* + \frac{\eta}{p}(2(C_* c_p + A_* + B_*)\|w\|_* + D_*)\right)\frac{1}{d^p(t,z)}$$

implying $W(t,z) \in W_*(M)$, i.e., the operator (8.57) maps $W_*(M)$ into itself.

Next we ask under which conditions the operator (8.57) is contractive. For this end we choose a second element $\tilde{w} = \tilde{w}(t,z)$ in addition to $w = w(t,z)$. Denote the image of $\tilde{w} = \tilde{w}(t,z)$ by $\tilde{W} = \tilde{W}(t,z)$, i.e.,

$$\tilde{W}(t,z) = w_0(z) + \int_0^t L\tilde{w}(\tau,z)\,d\tau .$$

Then it follows

$$W(t,z) - \tilde{W}(t,z) =$$
$$\int_0^t \left\{C(\tau,z)\left(\frac{\partial w}{\partial z} - \frac{\partial \tilde{w}}{\partial z}\right) + A(\tau,z)(w - \tilde{w}) + B(\tau,z)\overline{(w - \tilde{w})}\right\}d\tau . \tag{8.66}$$

Since $w - \tilde{w} \in W_*(M)$ we have

$$\|w(\tau,z) - \tilde{w}(\tau,z)\|_{s(z)} \leq \frac{\|w - \tilde{w}\|_*}{d^p(\tau,z)} .$$

Applying the Nagumo lemma to $w - \tilde{w}$, one obtains

$$\left\|\frac{\partial w}{\partial z}(\tau,z) - \frac{\partial \tilde{w}}{\partial z}(\tau,z)\right\|_{s(z)} \leq \frac{c_p}{d^{p+1}(\tau,z)}\|w - \tilde{w}\|_* .$$

Once more taking into consideration the assumptions (8.58) - (8.60), in the space $C^\alpha(\overline{G}_{s(z)})$ the norm of the integrand of (8.66) can be estimated by

$$2(C_* c_p + A_* + B_*)\frac{\|w - \tilde{w}\|_*}{d^{p+1}(\tau,z)} .$$

Again applying lemma 5 of 8.1. to (8.66), we get the estimate

$$\|W(t,z) - \tilde{W}(t,z)\|_{s(z)} \leq \frac{2\eta}{p}(C_* c_p + A_* + B_*)\frac{\|w - \tilde{w}\|_*}{d^p(t,z)} ,$$

i.e., the operator (8.57) is contractive if

$$\frac{2\eta}{p}(C_* c_p + A_* + B_*) < 1.$$

Consequently, in view of the contraction-mapping principle the following theorem has been proved:

<u>Theorem.</u> Suppose that $\tilde{L}$ and l are associated. Suppose, further, that the coefficients of $\tilde{L}$ and the initial function $w_0 = w_0(z)$ as well satisfy the assumptions (8.58) - (8.62) in a conical domain M, where $p > 0$ is any given number. Restrict η to

$$\eta < \frac{p}{2(C_* c_p + A_* + B_*)}$$

if necessary. Then the initial value problem (8.54), (8.55) can be solved globally in M by successive approximations. The solution is uniquely determined in $W_*(M)$.

Notice that the approcimations $w_k(t,z)$ can be represented in the following form:

Substituting $w(t,z) = 0$ into the right-hand side of (8.57), we get the first approximation $w_1(t,z)$, i.e.,

$$w_1(t,z) = w_0(z) + \int_0^t D(\tau,z)d\tau.$$

The (k+1)-th approximation is defined recursively by substituting $w_k(t,z)$ into the right-hand side of (8.57), i.e.,

$$w_{k+1}(t,z)$$
$$= w_0(z) + \int_0^t (C(\tau,z)\frac{\partial w_k}{\partial z} + A(\tau,z)w_k + B(\tau,z)\overline{w}_k + D(\tau,z))d\tau.$$

8.3. A weighted norm for functions depending on time in scales of Banach spaces

Now we are going to generalize the construction of the conical domain M (cf. 8.1., p. 132, and 8.2., p. 145) to the case of abstract scales of Banach spaces. For this end we regard an arbitrary scale of Banach spaces B_s, $0 < s < s_0$, and define M by

$$M = \{(t,s) : 0 < s < s_0, \ 0 \leq t < \eta(s_0 - s)\},$$

where η will be fixed later. In this case the non-negative value

$$d(t,s) := s_0 - s - \frac{t}{\eta}$$

may be interpreted as measure for the distance of (t,s) from the lateral surface of M. The intersection of M with the hyperplane $t = t'$ is

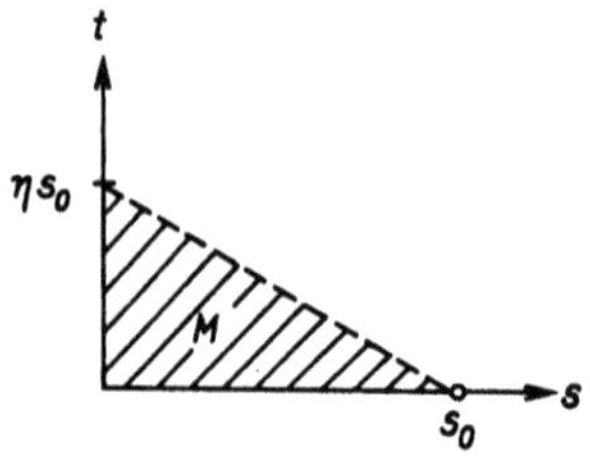

given by

$$\{(t',s) : s < s'\},$$

where s' is defined by

$$s' = s_0 - \frac{t'}{\eta} .$$

Further, take any δ, $0 < \delta < s_0$. Then

$$\tilde{K} = \{(t,s) : 0 < s < s_0, \quad 0 \leq t \leq \eta(s_0 - s - \delta)\} \quad (8.67)$$

is a subset of M which can be used instead of the corresponding compact subsets introduced in the sections 8.1. (see p. 133) and 8.2. (see p. 145). To each point (t,s) of M there exists at least one δ such that (t,s) belongs to the set $\tilde{K}$ defined by (8.67). Notice that

$$d(t,s) = s_0 - s - \frac{t}{\eta} \geq \delta \quad (8.68)$$

for every (t,s) belonging to $\tilde{K}$.

In the following we say that $u = u(t)$ is defined in M if u(t) belongs to B_s in case that $0 \leq t < \eta(s_0 - s)$. For functions $u = u(t)$ defined in M we regard the functional

$$\|u(t)\|_* = \sup_M \|u(t)\|_s (s_0 - s - \frac{t}{\eta})^p = \sup_M \|u(t)\|_s d^p(t,s), \quad (8.69)$$

where p is a fixedly chosen positive number. Denote the space of all continuous functions $u = u(t)$ for which $\|u(t)\|_*$ is finite by $B_*(M)$. Since

$$\|u_1(t) + u_2(t)\|_s \leq \|u_1(t)\|_s + \|u_2(t)\|_s$$

for each s and each t the functional

$$\|u_1 + u_2\|_*$$

is finite if $\|u_1\|_*$ and $\|u_2\|_*$ as well are finite. Thus $B_*(M)$ turns out to be a linear space. It is easily seen, further, that the functional $\|\cdot\|_*$ (8.69) is a norm in $B_*(M)$.

In order to prove that $B_*(M)$ is complete, we take any fundamental sequence $\{u_n(t,s)\}_{n=1,2,\ldots}$ in M. In view of (8.68) we have

$$\|u_n(t) - u_m(t)\|_s \leq \frac{\|u_n(t) - u_m(t)\|_*}{\delta^p} \quad (8.70)$$

for points (t,s) belonging to $\tilde{K}$, i.e., the $u_n(t)$ are a fundamental sequence in B_s provided $0 \leq t \leq \eta(s_0 - s - \delta)$. Since B_s is complete there exists a limit function $u = u(t)$. On the other hand, each point of M belongs at least to one $\tilde{K}$. Therefore, the limit function $u = u(t)$ is defined in the whole set M. In view of (8.70) the convergence is uniform in each $\tilde{K}$ and, consequently, the limit function depends continuously on t.

For sufficiently large n, m we have

$$\|u_n(t) - u_m(t)\|_s d^p(t,s) \leqq \|u_n - u_m\|_* < \varepsilon.$$

Carrying out the limiting process $m \to \infty$, one gets

$$\|u_n(t) - u(t)\|_s d^p(t,s) \leqq \varepsilon,$$

i.e.,

$$\|u_n(t) - u(t)\|_* = \sup_M \|u_n(t) - u(t)\|_s d^p(t,s) \leqq \varepsilon$$

for sufficiently large n. Therefore, the $u_n(t)$ converge to u(t) with respect to the metric of $B_*(M)$. This proves the completeness of that space.

Now regard again the abstract initial value problem

$$\frac{du}{dt} = F(t,u), \tag{3.2}$$

$$u(0) = u_o, \tag{3.3}$$

where the right-hand side F(t,u) maps the scale B_s into itself. We look for solutions belonging to the space $B_*(M)$.

Take into account that the definition (8.70) of the norm $\|\cdot\|_*$ does not exclude that $\|u(t)\|_s$ is unbounded near the boundary of M because the factor $d^p(t,s)$ tends to zero as (t,s) tends to the lateral surface of M (if $\|u(t)\|_s$ does not tend faster to $+\infty$ than $d^p(t,s)$ goes to zero, then the right-hand side of (8.69) remains bounded, i.e., u = u(t) belongs to $B_*(M)$). Therefore, the space $B_*(M)$ is most useful for solving the initial value problem (3.2), (3.3) in case that the right-hand side F(t,u) of (3.2) is defined for all u. Therefore, we assume that the right-hand side F(t,u) is a linear operator satisfying the conditions (I'), (II), (III) (cf. 3.8. and 3.2.; cf. also the remark on the linearity of F(t,u) in section 3.8., p. 50).

The initial value problem (3.2), (3.3) is equivalent to the integral equation

$$u(t) = u_o + \int_0^t d\tau \cdot F(\tau,u(\tau)) \tag{3.17}$$

(see section 3.3.). Therefore, we have to look for fixed points of the operator defined by

$$U(t) = u_o + \int_0^t d\tau \cdot F(\tau,u(\tau)). \tag{8.71}$$

Take any element u = u(t) belonging to $B_*(M)$. Choose, further, any point

(t,s) in M. Then define $\tilde{s}$ by

$$\tilde{s} = s + \frac{1}{p+1} d(t,\tau), \tag{8.72}$$

where

$$d(\tau,s) = s_0 - s - \frac{\tau}{\eta}$$

and $0 \leq \tau \leq t$. Since

$$d(\tau,\tilde{s}) = s_0 - \tilde{s} - \frac{\tau}{\eta} = s_0 - s - \frac{\tau}{\eta} - \frac{1}{p+1} d(t,\tau) = \frac{p}{p+1} d(t,\tau) > 0$$

the pair $(\tau,\tilde{s})$ belongs to M, too, and $u = u(\tau)$ belongs to $B_{\tilde{s}}$. This implies that $F(\tau,u(\tau))$ belongs to B_s.

Suppose, moreover, that the initial value u_0 can be estimated by

$$\| u_0 \|_s \leq \frac{\text{const}}{(s_0 - s)^p} . \tag{8.73}$$

Then

$$\| u_0 \|_s d^p(t,s) \leq \text{const} \left(\frac{s_0 - s - \frac{t}{\eta}}{s_0 - s} \right)^p \leq \text{const},$$

i.e., u_0 belongs to $B_*(M)$, too.

In order to prove that the image $U(t)$ defined by (8.71) is also an element of $B_*(M)$, it remains to show that the integral in (8.71) belongs to $B_*(M)$. For this end we rewrite $F(\tau,u(\tau))$ in the form

$$F(\tau,u(\tau)) = (F(\tau,u(\tau)) - F(\tau,u_0)) + F(\tau,u_0).$$

Using the properties (II) and (III) of $F(t,u)$, we get

$$\| F(\tau,u(\tau)) \|_s \leq \| F(\tau,u(\tau)) - F(\tau,u_0) \|_s + \| F(\tau,u_0) \|_s$$
$$\leq \frac{C}{\tilde{s} - s} \| u(\tau) - u_0 \|_{\tilde{s}} + \frac{K}{s_0 - s} .$$

Once more taking into account the definition (8.69) of the $\| \cdot \|_*$-norm, we get

$$\| u(\tau) - u_0 \|_{\tilde{s}} \leq \frac{\| u(\tau) - u_0 \|_*}{d^p(\tau,\tilde{s})} = \frac{\| u(\tau) - u_0 \|_*}{d^p(\tau,s)} (1 + \frac{1}{p})^p .$$

Therefore, one has

$$\| F(\tau,u(\tau)) \|_s \leq C(p+1)(1 + \frac{1}{p})^p \frac{\| u(\tau) - u_0 \|_*}{d^{p+1}(\tau,s)} + \frac{K}{s_0 - s} .$$

Applying lemma 5 of 8.1., we see that

$$\int_0^t \| F(\tau,u(\tau)) \|_s d\tau \leq C\eta(1 + \frac{1}{p})^{p+1} \frac{\| u(\tau) - u_0 \|_*}{d^p(t,s)} + \frac{Kt}{s_0 - s} .$$

Since

$$0 \leq t < \eta(s_0 - s)$$

and

$$d(t,s) = s_0 - s - \frac{t}{\eta} \leq s_0$$

the second term on the right-hand side of the last inequality can be estimated by

$$K\eta \leq K\eta \frac{s_0^p}{d^p(t,s)} .$$

Hence it follows that

$$\|U(t) - u_0\|_s \leq \eta(C(1 + \frac{1}{p})^{p+1}\|u(\tau) - u_0\|_* + Ks_0^p) \frac{1}{d^p(t,s)}$$

and, consequently, $\|U(t) - u_0\|_*$ is finite, i.e., $U(t) - u_0$ belongs to $B_*(M)$. Since $u_0 \in B_*(M)$ one obtains, finally, that $U(t) \in B_*(M)$, i.e., the operator (8.71) maps $B_*(M)$ into itself.

It remains to look for sufficient conditions under which the operator (8.71) turns out to be contractive. For this take two elements u(t) and v(t) belonging to $B_*(M)$ and regard the images U(t) and V(t) resp. defined by (8.71) and

$$V(t) = u_0 + \int_0^t d\tau \cdot F(\tau, v(\tau)),$$

respectively. Then we get

$$\|U(t) - V(t)\|_s \leq \int_0^t \|F(\tau,u(\tau)) - F(\tau,v(\tau))\|_s d\tau$$

$$\leq \frac{C}{\tilde{s} - s} \int_0^t \|u(\tau) - v(\tau)\|_{\tilde{s}} d\tau,$$

where $\tilde{s}$ is defined by (8.72). Again applying the definition (8.69), one gets analogously

$$\|u(\tau) - v(\tau)\|_{\tilde{s}} \leq \frac{\|u(\tau) - v(\tau)\|_*}{d^p(\tau,\tilde{s})} = \frac{\|u(\tau) - v(\tau)\|_*}{d^p(\tau,s)}(1 + \frac{1}{p})^p$$

and, finally,

$$\|U(\tau) - V(\tau)\|_s \leq C(p + 1)(1 + \frac{1}{p})^p\|u(\tau) - v(\tau)\|_* \int_0^t \frac{d\tau}{d^{p+1}(\tau,s)}$$

$$\leq C\eta(1 + \frac{1}{p})^{p+1}\|u(\tau) - v(\tau)\|_* \cdot \frac{1}{d^p(t,s)} .$$

This implies

$$\|U(\tau) - V(\tau)\|_* = \sup_M \|U(\tau) - V(\tau)\|_s d^p(t,s)$$

$$\leq C\eta(1 + \frac{1}{p})^{p+1}\|u(\tau) - v(\tau)\|_* ,$$

i.e., the operator (8.71) is contractive provided

$$\eta < \frac{1}{C(1 + \frac{1}{p})^{p+1}} . \qquad (8.74)$$

Summarizing these arguments we see that the following theorem is true:

<u>Theorem.</u> Assume the right-hand side F(t,u) of the differential equa-

tion (3.2), (3.3) satisfies the conditions (I'), (II), and (III). Assume, further, that the initial function u_0 can be estimated by (8.73). Then there exists a unique solution of the initial value problem (3.2), (3.3) belonging to $B_*(M)$ provided η satisfies the condition (8.74).

Concluding our considerations on contraction-mapping principles for initial value problems in scales of Banach spaces, we would like to remark that similar constructions are possible even in the case that $F(t,u)$ is defined only in a ball centred at u_0. Replacing the norm (8.69) by

$$\|u(t)\|_* = \sup_{0<s<s_0} \sup_{0<t<\eta(s_0-s)} \|u(t)\|_s \left(\frac{\eta(s_0 - s)}{t} - 1\right), \tag{8.75}$$

a contraction-mapping principle has been applied for solving the initial value problem (3.2), (3.3) in the paper [73]. Notice, finally, that the norm (8.75) is identical with the functional $M(u)$ introduced in section 3.3. if the parameter η in the definition of $\|\cdot\|_*$ is denoted by a.

9. FURTHER EXISTENCE THEOREMS FOR INITIAL VALUE PROBLEMS IN SCALES OF BANACH SPACES

In the chapters 7 and 8 we have solved initial value problems with generalized analytic initial functions. The solution of still more general initial value problems requires the modification of the scale used in chapter 7. In the present chapter we would like to point to some further applications of the method of scales of Banach spaces and related methods in the theory of partial differential equations.

Contrary to the preceding chapters 1 - 5 and 7 - 8, the present chapter 9 as well as the following chapter 10 do not contain complete proofs, but some basic ideas will be explained in detail.

References to papers on adjoining problems will be given.

9.1. Scales of q-holomorphic and generalized q-holomorphic vectors

9.1.1. First note the important fact that the differential equation (6.47) may be regarded as canonical form of a (homogeneous) linear and uniformly elliptic first order system for two desired real-valued functions u and v in the plane. Transforming the independent variables x, y and replacing the unknown functions $u = u(x,y)$, $v = v(x,y)$ by linear combinations of the originally sought ones, a given uniformly elliptic system can be rewritten in the form (6.47) (see, for instance, chapter 2 of I. N. Vekua's book [77]; cf. also section 6.3.5.).

In the case of systems of 2n linear first order equations for 2n unknown real-valued functions $u_1, \ldots, u_n, v_1, \ldots, v_n$ in the plane we introduce vectors $w = (w_1, \ldots, w_n)$ with complex-valued components $w_j = u_j + iv_j$, $j = 1, \ldots, n$. Then the given system may be rewritten as vector equation containing both derivatives $\partial w/\partial z$ and $\partial w/\partial \bar{z}$. As mentioned above in the case $n = 1$ one of the two derivatives $\partial w/\partial z$ and $\partial w/\partial \bar{z}$ may be eliminated and equation (6.47) may be interpreted as canonical form of a given (homogeneous) system. If $n \geq 2$ then it is not possible to eliminate one of the derivatives $\partial w/\partial z$ and $\partial w/\partial \bar{z}$, in general. Following B. Bojarski [9], homogeneous uniformly elliptic systems may be rewritten, however, in the form

$$\frac{\partial w}{\partial \bar{z}} - q(z)\frac{\partial w}{\partial z} - a(z)w - b(z)\bar{w} = 0, \tag{9.1}$$

where the coefficients are quasidiagonal matrices. Solutions to this equation are called <u>generalized q-holomorphic vectors</u>, whereas <u>q-holomorphic vectors</u> are solutions to the special equation

$$\frac{\partial w}{\partial \bar{z}} - q(z)\frac{\partial w}{\partial z} = 0. \tag{9.2}$$

Systems for 2n desired real-valued functions may be also rewritten by using Douglis' concept of hypercomplex variables. Regarding this approach to the theory of linear systems for 2n sought functions we refer to R. P. Gilbert's book [25].

Using scales of Banach spaces of both q-holomorphic and generalized q-holomorphic vectors, A. Crodel [14, 15] solved initial value problems for differential operators of type

$$\frac{\partial w}{\partial t} = C(t,z)\frac{\partial w}{\partial z} + A(t,z)w + B(t,z)\bar{w} + D(t,z), \tag{9.3}$$

where the coefficients are matrices provided the right-hand side of the differential equation (9.3) is associated to the operator on the right-hand side of (9.1) and (9.2), respectively. In the case of generalized q-holomorphic vectors the corresponding Banach spaces must be equipped either with a Hölder-norm (cf. 6.2.6.) or an L_p-norm, $p > 2$ (cf. (6.45)), whereas the scales of Banach spaces of q-holomorphic functions may be equipped with the supremum norm (cf. (2.1)), in a similar way as done in the holomorphic case of chapter 4 (take into consideration that the definition (2.1) as well as the definitions (6.45) and 6.2.6. are to modify since $w = w(z)$ is a vector in the present section). Using results of B. Bojarski [9] and B. Goldschmidt [26], A. Crodel proved that the operators defined by the right-hand side of (9.3) are generalized Cauchy-Riemann operators. Thus the initial value problems under consideration may be solved by applying the methods of chapter 3. In the linear cases (9.3), (9.2) and (9.3), (9.1) associated differential equations are constructed in A. Crodel's papers [16, 17].

Applying A. Crodel's a-priori estimates of the derivative of a q-holomorphic and a generalized q-holomorphic vector resp., initial value problems for (9.3) can also be solved by using a modified weighted supremum norm in the sense of section 8.2. (cf. also [74]).

H. Begehr [6] solves initial value problems with hyperanalytic initial functions.

9.1.2. In his paper [14] A. Crodel solved also initial value problems for quasilinear equations of type

$$\frac{\partial w}{\partial t} = C(t,z,w)\frac{\partial w}{\partial z} + D(t,z,w), \qquad (9.4)$$

where the desired solution is assumed to be q-holomorphic at each t. In this paper the initial value problem is investigated even for some kinds of right-hand sides depending nonlinearly on the derivative of the vector looked for.

A. Crodel [15] constructs also solutions of the initial value problem for (9.4) in classes of generalized q-holomorphic vectors. In his paper [18] A. Crodel finds nonlinear right-hand sides of type (9.4) associated to (9.1) provided the rank of b(z) is at most equal to n - 1.

9.2. Scales of pseudoholomorphic functions in L. Bers' sense

Generalized analytic functions in I. N. Vekua's sense are a generalization of holomorphic functions. The starting point of this theory is replacing the differential equation (6.13) for holomorphic functions by the more general equation (6.47). Another possibility of generalizing holomorphic functions is the following: one replaces the complex difference quotient (with the help of which the complex derivative is defined, cf. 6.1.5.) by a more general one containing two auxiliary functions F = F(z), G = G(z). Functions w = w(z) for which this generalized difference quotient exists at each point of a given domain G are called (F,G)-pseudoholomorphic in G (see L. Bers [7]). In the paper [75] scales of Banach spaces of pseudoholomorphic functions have been constructed. These scales enable us to solve initial value problems with pseudoholomorphic initial functions. The mentioned paper [75] contains, as a side result, a modification of L. Bers' (F,G)-derivative possessing the following property: the modified differential operator maps the space of pseudoholomorphic functions into itself.

9.3. Commentary on connections between the Cauchy-Kovalevskaya theorem and other problems in Mathematical Analysis

9.3.1. The above results show that generalizations of the Cauchy-Kovalevskaya theorem are possible if holomorphic functions are replaced by generalized analytic functions (cf. 7.4., 8.2.), q-holomorphic and generalized q-holomorphic vectors (cf. 9.1.), and by pseudoholomorphic functions (cf. 9.2.). It is natural to ask whether sufficiently strong differentiability properties are also sufficient for the solvability of initial value problems for first order differential equations with complex-valued initial functions. The answer is no because there are even linear differential equations with C^{∞}-coefficients without any solutions (see, however, also section 9.3.3.). The first example of a differential equation like that was given by H. Lewy. In his paper [35] (see also F. John's book [29]) he constructed a function $F = F(x,y,t)$ for which the differential equation

$$\frac{\partial u}{\partial x} + i\frac{\partial u}{\partial y} - 2i(x + iy)\frac{\partial u}{\partial t} = F(x,y,t)$$

does not have any solution. This example shows that rigorous weakenings of the assumptions on the right-hand sides of the differential equations with respect to the spacelike variables cannot be expected.

The L. Lewy example was the starting point for the development of a special field within the theory of partial differential equations aimed at the formulation of the conditions under which a given equation does possess at least one solution. General results in this direction were obtained, for instance, by L. Hörmander [28] and Yu. V. Egorov. Readers who are interested in the history of this field should carefully read the introduction written by O. A. Oleinik, S. L. Sobolev, and A. N. Tikhonov to the proceedings [3] and also Yu. V. Egorov's paper in the same proceedings.

In constrast to that, the present book is aimed at deducing such conditions under which even the initial value problem is solvable. The result of chapter 3 is an abstract version of sufficient conditions for the solvability of the last problem. A concrete version of these conditions is, for instance, that the initial function is holomorphic and further that the right-hand sides of the differential equations transform holomorphic functions into themselves (cf. chapter 4). The sections 7.4. and 8.2. show that the initial function need not be holomorphic but it is sufficient that it is a generalized analytic function (at the same time the assumption on the right-hand side of the differential equation must be replaced by the following one: the right-hand side is a differential operator associated to the differential equation for the initial function). The considerations 9.4. will show that a suitable modification of the concept of holomorphy yields sufficient conditions for the solvability of initial value problems in the case of more than two real spacelike variables.

9.3.2. The classical proof of the Cauchy-Kovalevskaya theorem makes use of power series representations of the desired solutions in all variables. This implies that the right-hand side of the differential equation (4.2) has to be power series in all variables.

Nagumo's approach to the Cauchy-Kovalevskaya theorem was based on the so-called Nagumo lemma (lemma 1 of 8.1.; the estimate (2.2) is the special case $p = 0$ of this lemma). This approach led not only to the classical Cauchy-Kovalevskaya theorem, but it gave a positive answer to the question:

Does there exist a solution to the initial value problem (4.2), (4.5) under the assumption that the right-hand sides depend only continuously on t, whereas they possess power series representation with respect to the remaining variables?

A positive answer to this question was given by M. Nagumo in his paper [44]. T. Yamanaka [80] solved this problem by using scales of Banach spaces. His approach includes, however, the linear case only. An abstract version of the non-linear Cauchy-Kovalevskaya theorem applicable to the general non-linear case was only given by T. Nirenberg [45]. Concerning historical remarks about the problem under consideration we would like to hint also to the supplement of the Russian translation of T. Nirenberg's book [46].

A. A. Agrachev and S. A. Vakhramev regard in their paper [1] the case that the right-hand side is only a measurable and locally integrable function in t.

9.3.3. Let us recall that the classical proof of the Cauchy-Kovalevskaya theorem is based on power series representations, i.e., the desired solution is written as a power series (cf. also 9.3.2.). The coefficients are calculated recursively by substituting this formal series into the differential equation and the initial condition as well. The convergence of the series can be proved by the comparison method.

It remains to investigate whether the formal power series can converge also under the assumption that the right-hand side of the differential equation or the initial function are not power series but only infinitely differentiable functions.

This can be done if the right-hand sides and the initial functions belong to so-called Gevrey classes, i.e., they are infinitely differentiable and their derivatives can be estimated in the following way:

The function $f = f(x_1,\dots,x_n)$ belongs to the class $(\delta_1,\dots,\delta_n)$ if

$$\left|\frac{\partial^{k_1+\dots+k_n} f}{\partial x_1^{k_1}\dots\partial x_n^{k_n}}\right| \leqq C_0 C^{k_1+\dots+k_n}(\delta_1 k_1)! \;\dots\; (\delta_n k_n)!,$$

where $(\delta k)!$ is defined by the Γ-function,

$$(\delta k)! = \Gamma(\delta k+1).$$

A first result in this direction was obtained by Le Roux [34]. He solved the initial value problem

$$\frac{\partial^2 u}{\partial t^2} = \frac{\partial u}{\partial x},$$

$$u(0,x) = \phi_0(x),$$

$$\frac{\partial u}{\partial t}(0,x) = \phi_1(x),$$

where ϕ_0, ϕ_1 belong to a Gevrey class. Non-linear equations have been investigated by V. R. Friedlender [22] and A. Friedman [23].

9.3.4. The method of scales of Banach spaces as well as contraction-mapping principles in the sense of section 8.1. can also be applied for solving initial value problems for functional differential equations containing terms such as

$$u(t-\Theta),$$

where Θ is a positive number, $0 \leqq \Theta \leqq \Theta_0$. In this case the initial function must be prescribed not only for $t = 0$ but also in the whole interval $-\Theta_0 \leqq t \leqq 0$.

For details we refer to the papers of T. Sekine and T. Yamanaka [58], of T. Yamanaka and H. Tamaki [82], and of W. Walter [79].

9.3.5. Other generalizations of the Cauchy-Kovalevskaya theorem deal with initial value problems for pseudodifferential operators. In M. S. Baouendi's and C. Goulaounic's paper [4] such generalization of the non-linear Cauchy-Kovalevskaya theorem and its application to the Euler equation for nonviscous incompressible fluids is given.

9.3.6. Although the Cauchy-Kovalevskaya theorem is a central and autonomous result in mathematics, it can be deduced from other general theorems such as the contraction-mapping principle (see 8.1.). In K. Keller's and A. Schneider's paper [30] the proof of the Cauchy-Kovalevskaya theorem is based on the Schauder-Tikhonov fixed-point theorem. It may be added that Schauder's fixed-point theorem is used also in M. Nagumo's fundamental paper [44], whereas in H. Okamura's paper [48] the Nagumo approach is carried out by using successive approximations.

9.3.7. On the right-hand side of the m-th order Cauchy-Kovalevskaya differential equations (4.2) only derivatives up to the order m are permitted, still higher derivatives must not occur. This condition is not only sufficient but also necessary. This has been proved by S. Mizohata in his paper [40] generalizing his previous one [39] in which the case of coefficients depending only on the time has been investigated.

9.3.8. The development of the method of scales of Banach spaces has been closely connected not only with an abstract version of the method of successive approximations but also with uniqueness theorems for the Cauchy problem for partial differential equations. In this respect we refer to I. M. Gelfand's and G. E. Shilov's book [24]. There the initial value problem

$$\frac{du}{dt} = Au, \tag{9.5}$$

$$u(0) = u_o \tag{9.6}$$

(see (3.2), (3.3)) is solved for a linear operator not depending on t and mapping a scale B_s, $0 < s < s_o$, into itself. Suppose that the norm of A interpreted as an operator mapping B_s into $B_{s'}$, $0 < s' < s < s_o$, can be estimated by

$$\|A\|_{s'}^{s} \leq \frac{C}{s - s'} \tag{9.7}$$

(cf. (3.16)), where $\|A\|_{s'}^{s}$ means the norm of A as an operator mapping B_s into $B_{s'}$. For solving the initial value problem (9.5), (9.6) the operator exp(tA) defined by

$$\sum_{n=o}^{\infty} \frac{t^n}{n!} A^n \tag{9.8}$$

can be used. The proof of the convergence of the series (9.7) is based on the following estimate:

Subdividing the interval $[s',s]$ into n subintervals $[s_{i-1},s_i]$, $i = 1, \dots, n$, of equal length

$$s_i - s_{i-1} = \frac{s - s'}{n},$$

we get

$$\|A^n\|_{s'}^{s} \leq \prod_{i=1}^{n} \|A\|_{s_{i-1}}^{s_i}.$$

In view of (9.7) one has

$$\|A\|_{s_{i-1}}^{s_i} \leq \frac{C}{s_i - s_{i-1}} = \frac{Cn}{s - s'}$$

and, consequently,

$$\|A^n\|_{s'}^{s} \leqq \left(\frac{Cn}{s - s'}\right)^n .$$

The quotient of the norms of two successive terms of the series (9.8) can be estimated by

$$\frac{t}{n+1} \frac{C}{s-s'} \frac{(n+1)^{n+1}}{n^n} = \frac{tC}{s-s'} (1 + \frac{1}{n})^n \leqq \frac{tCe}{s-s'} \tag{9.9}$$

Therefore, the series converges provided

$$\frac{tCe}{s-s'} < 1.$$

In his paper T. Yamanaka [80] generalizes this construction to the case that the operator A depends on t. Notice that the estimate (3.77) is another version of the estimate (9.9).

9.4. Scales of Banach spaces in the case of more than 2 spacelike variables

9.4.1. The above considerations on scales of Banach spaces of generalized analytic functions and their application to initial value problems can be generalized to the case of more than two real spacelike variables. The simplest possibility to generalize the concept of holomorphic functions to the case of the three-dimensional Euclidian space $\mathbb{R}^3$ is the following: The Cauchy-Riemann system (6.11) is to be replaced by the system

$$\begin{aligned} \operatorname{div} u &= 0, \\ \operatorname{rot} u &= 0, \end{aligned} \tag{9.10}$$

where $u = (u_1, u_2, u_3)$ is the desired vector depending on $x = (x_1, x_2, x_3)$. The solutions u to (9.10) are called potential vectors. Now we look for first order differential operators of type

$$Lu = \left\{\sum_{i,j} A^1_{ij} \frac{\partial u_i}{\partial x_j} + \sum_i B^1_i u_i + D^1, \ldots\right\}, \tag{9.11}$$

where the coefficients A^K_{ij}, B^K_i, D^K depend on both the time t and the spacelike variable x. In the paper [66] sufficient conditions under which L transforms the space of all potential vectors into itself have been deduced. Such operators are associated to the left-hand sides of (9.10) in the sense of the definition 6.4.1. In the paper [68] it has been proved, for instance, that 9 out of the 27 coefficients A^K_{ij} of L may be chosen arbitrarily. For such operators L the initial value problem

$$\begin{aligned} \frac{\partial u}{\partial t} &= Lu, \\ u(0) &= u_0, \end{aligned} \tag{9.12}$$

where the initial value u_o is a potential vector is also solvable by the method of scales of Banach spaces.

In the paper [69] the system (9.11) is replaced by the more general one

$$\begin{aligned} \operatorname{div} u + (a, u) &= 0, \\ \operatorname{rot} u + u \times b &= 0, \end{aligned}$$

where a and b are given constant vectors (its solutions are called <u>generalized potential vectors</u>).

<u>9.4.2.</u> In order to generalize the concept of holomorphy in the case of still more real variables it is useful to introduce the concept of <u>monogenic functions</u> (cf. the book of F. Brackx, R. Delanghe, and F. Sommen [10]). For this end we first introduce a Clifford algebra $\mathcal{A}$ generated by $e_o, e_1, \ldots, e_n$, where $e_o = 1$. $e_\alpha^2 = -1$, $e_\alpha e_\beta + e_\beta e_\alpha = 0$, $e_\alpha(e_\beta e_\gamma) = (e_\alpha e_\beta)e_\gamma$ for $\alpha, \beta, \gamma = 1, \ldots, n$. Note that in the case $n = 1$ the element e_1 is usually denoted by i. Next define the differential operator D by setting

$$D = \sum_{i=0}^{n} e_i \frac{\partial}{\partial x_i} .$$

Then an $\mathcal{A}$-valued function u defined in a given domain G of $\mathbb{R}^{n+1}$ is said to be left-monogenic if it satisfies the differential equation

$$Du = 0$$

everywhere in G. Notice, further, that in the case $n = 1$ the last differential equation is identical with the complex form (6.13) of the Cauchy-Riemann system in the plane (multiplied with the factor 2). Initial value problems of type (9.12), where L is a linear first order operator acting in $\mathbb{R}^{n+1}$ are solvable by the method of scales of Banach spaces provided both L transforms monogenic functions into themselves and the initial function u_o is monogenic (see [70]).

9.5. The Ovsyannikov scale

The scales of which we made use in the chapters 4 and 7 are defined as solutions to given differential equations in a family of subdomains G_s of a given domain (remember that also holomorphic functions in several complex variables may be described by a system of partial differential equations, cf. 6.1.8.). In these cases the parameter s of the Banach spaces B_s that form the scale is related to the parameter s of the family of subdomains G_s (remember further that in chapter 4 the Banach spaces are denoted by H_s whereas in chapter 7 they are denoted by W_s). The scale parameter s, however, need not necessarily be related to a family of subdomains. The following Ovsyannikov scale set an example for a concrete scale for which the scale parameter is not related to a family of subdomains.

In order to introduce the Osyannikov scale we take any compact subset K of the n-dimensional Euclidian space $\mathbb{R}^n$ whose points are denoted by $x = (x_1, \ldots, x_n)$. Denote the differential operator

$$\frac{\partial^k}{\partial x_1^{\alpha_1} \ldots \partial x_n^{\alpha_n}}$$

by D, where $\alpha = (\alpha_1, \ldots, \alpha_n)$ and $|\alpha| = \alpha_1 + \ldots + \alpha_n = k$. Let further $u = u(x)$ be any real-valued infinitely differentiable function defined on the whole space $\mathbb{R}^n$. Now take any real number s, $0 < s < 1$. Denote by B_s the set of all those functions $u = u(x)$ for which the norm $\|u\|_s$ defined by

$$\|u\|_s = \sum_{k=0}^{\infty} \frac{s^k}{k!} \max_{|\alpha|=k} \sup_K |D^\alpha u| \tag{9.13}$$

is finite. Then the B_s, $0 < s < 1$, form a scale of Banach spaces. From the definition (9.13) it follows immediately that

$$\|u\|_{s'} \leqq \|u\|_s$$

if $s' < s$. The special norm (9.13) possesses, however, some further properties, for instance

$$\|uv\|_s \leqq \|u\|_s \cdot \|v\|_s .$$

The Ovsyannikov scale, equipped with the norm (9.13), can be used for solving free boundary problems in hydromechanics (see L. V. Ovsyannikov's paper [51] in which also further references to applications of scales in mechanics are given). A new application of L. V. Ovsyannikov's method in hydrodynamics is contained in J. Duchon's and R. Robert's paper [21].

9.6. Solution of initial value problems in scales of Banach spaces by Euler's polygonal line method

Instead of the method of successive approximations (cf. chapter 3) also Euler's polygonal line method may be applied for solving initial value problems for ordinary differential equations in scales of Banach spaces (see H. Begehr's paper [5]). For applying this method we have to suppose, however, that the scale B_s, $0 < s < s_0$, under consideration possesses the following additional property:

For any pair s, s' with $0 < s' < s < s_0$ the operator $I_{s,s'}$ imbedding B_s into $B_{s'}$ is compact.

In view of Arzelà's theorem the scales of Banach spaces of holomorphic functions introduced in section 2.2. satisfy this condition because uniformly bounded holomorphic functions are equicontinuous in compact subsets of their domain of definition.

Since in view of (6.55) generalized analytic functions may be represented by holomorphic ones the same statement is true for scales of Banach spaces of generalized analytic functions. Finally the Ovsyannikov scale (cf. section 8.1.5.) satisfies the above condition (it may be added scales satisfying this condition are called K-scales, see L. V. Ovsyannikov [51]).

The following version of the Arzelà theorem holds for scales with compact imbedding operators:

<u>Lemma.</u> Suppose each $u_k = u_k(t)$, $k = 1, 2, \ldots$, maps the segment $[0, \tilde{T}]$ into B_s. Suppose, further, that the following two conditions are satisfied:

a) The norms $\| u_k(t) \|_s$, $k = 1, 2, \ldots$ and $0 \leq t \leq \tilde{T}$, are uniformly bounded.

b) The $u_k = u_k(t)$, $k = 1, 2, \ldots$, are equicontinuous at each point of $[0, \tilde{T}]$.

Then there exists a subsequence $\{u_{k_i}(t)\}_{i=1,2,\ldots}$ uniformly converging in $[0, \tilde{T}]$ with respect to $B_{s'}$, where $s' < s$.

This lemma may be proved in the same way as the usual version of the Arzelà theorem, i.e., we first choose a sequence of points $t_1, t_2, \ldots$ which are dense in $[0, \tilde{T}]$. Since a ball in B_s is compact in $B_{s'}$, the sequence $\{u_k(t_1)\}_{k=1,2,\ldots}$ contains a convergent subsequence. Next take a subsequence of this last one converging in t_2, too, and so on. The desired uniformly convergent sequence, finally, may be obtained by Cantor's diagonal procedure.

Now regard again the initial value problem (3.2), (3.3), where the right-hand side $F(t,u)$ is defined in $[0, T]$. We look for a solution $u = u(t)$ defined in $[0, \tilde{T}]$, where $\tilde{T} \leq T$. After subdividing the segment $[0, \tilde{T}]$ into n subintervals $[t_j, t_{j+1}]$, $j = 0, 1, \ldots, n-1$, where $t_o = 0$ and $t_n = T$ we define the piecewise linear function $\tilde{u} = \tilde{u}(t)$ in t_j, t_{j+1} by

$$\tilde{u}(t) = \tilde{u}(t_j) + (t - t_j)F(t_j, \tilde{u}(t_j)) \tag{9.14}$$

and we set

$$u(0) = u_o. \tag{9.15}$$

Now regard a sequence of subdivisions of $[0, T]$, where the length of the greatest subsegment of the k-th subdivision approaches zero as k approaches infinity. Denote the above-defined piecewise linear function that corresponds to the k-th subdivision by $\tilde{u}_k$. Then we look for conditions under which the u_k converge to a solution of the initial value problem (3.2), (3.3) as k tends to infinity.

Note that we meet with difficulties if we carry out this construction in the case of a general scale because $F(t_j,\tilde{u}(t_j))$ belongs only to $B_{s'}$ with $s' < s$ provided $u(t_j)$ belongs to B_s. The greater the number of subsegments of a subdivision, the more frequently we have to reduce the scale index s. On the other hand each transition of B_s to $B_{s'}$ is connected with the factor $\frac{1}{s - s'}$. Thus the estimates of the norms are worsening if the number of subsegments is increasing.

These difficulties are avoidable if we suppose that the right-hand side F(t,u) of the differential equation (3.2) does not reduce the domain of definition.

This condition is satisfied, for instance, if the scale is defined by solutions to partial differential equations (see the chapters 4 and 7). E.g., the complex differentiation $\frac{d}{dz}$ transforms holomorphic functions into holomorphic functions defined in the same domain. Now as ever we estimate the derivatives in subdomains (see section 2.3., formula (4.26), and the theorem in section 6.3.9.), but additionally we shall make use of the fact that the derivatives do exist in the whole domain. Strictly speaking we assume that the following conditions are satisfied:

a) The elements of B_s are functions defined in $\overline{G}_s$, where the G_s are subdomains of a given domain G having the properties formulated in section 2.2.

b) There exists a linear function space E whose elements are functions defined in G.

c) The restircitions of elements of E to $\overline{G}_s$ belong to B_s.

d) The initial function u_o is an element of E.

e) The right-hand side $F(t,\cdot)$ of the differential equation (3.2) transforms E into itself for each t.

These conditions are satisfied, for instance, in the cases investigated in the chapters 4 and 7. In the case of chapter 4 the space E contains all holomorphic functions in G, whereas in the case of chapter 7 the elements of E satisfy the associated differential equation in G (cf. section 7.4.).

Since E is a linear space all piecewise linear functions $\tilde{u}_k = \tilde{u}_k(t)$ map [0, T] into E. Thus for each t the $\tilde{u}_k(t)$ are defined in the whole domain

G and their restrictions to $\bar{G}_s$ belong to B_s. Our aim is to prove that the $\tilde{u}_k$ converge to a solution of the initial value problem (3.2), (3.3).

It is clear that a proof like this can be carried out only under suitable assumptions on the right-hand side F(t,u). In the following we outline a slightly modified version of H. Begehr's considerations [5]. H. Begehr supposes that F(t,u) satisfies the following two conditions instead of the conditions (I), (II), and (III) of section 3.2.:

(I') The right-hand side F(t,u) defines a continuous mapping of

$$\{t : 0 \leqq t \leqq T\} \times \{u \in B_s : \|u - u_0\|_s \leqq \frac{R}{s - s_0}\}$$

into $B_{s'}$, $s' < s$, where T and R are given numbers the first of which is identical with the one introduced at the beginning of the present section.

(II') The s'-norm of F(t,u) may be estimated by

$$\|F(t,u)\|_{s'} \leqq \frac{C}{s - s'}$$

if only $0 \leqq t \leqq T$ and $\|u - u_0\|_s \leqq \frac{R}{s - s_0}$.

Let us additionally remark that in the paper [5] the denominators $(s - s_0)$ and $(s - s')$ are replaced by $(s - s_0)^\gamma$ and $(s - s')^\gamma$ resp., where γ is any non-negative real number.

Since the definition (8.10) implies

$$\|\tilde{u}_k(t) - \tilde{u}_k(t_j)\|_{s'} \leqq (t - t_j)\|F(t_j, \tilde{u}_k(t_j))\|_{s'}$$
$$\leqq (t - t_j) \frac{C}{s - s'} \leqq (t - t_j)\frac{2C}{s_0 - s'}$$

for $t_j \leqq t \leqq t_{j+1}$ and $s = s' + \frac{1}{2}(s_0 - s)$ it follows by induction that

$$\|\tilde{u}_k(t) - u_0\|_s \leqq \frac{2CT}{s_0 - s} \leqq \frac{R}{s_0 - s}$$

provided $0 \leqq t \leqq \tilde{T}$ and

$$\tilde{T} \leqq \min (T, \frac{R}{2C}). \tag{9.16}$$

The last condition ensures, therefore, that all u_k may be defined recursively. Similarly one gets

$$\|\tilde{u}_k(t) - \tilde{u}_k(t')\|_s \leqq \frac{2C}{s_0 - s}|t - t'|.$$

Thus the u_k turn out to be equicontinuous mappings of $[0, \tilde{T}]$ into B_s.

Now take any monotonically increasing sequence $s_1, s_2, \ldots, s_1 < s_2 < \ldots$, approaching s_0. Since the s_2-norms of the u_k are uniformly bounded, in view of the lemma there exists a subsequence uniformly converging with respect to B_{s_1}. On the other hand, the s_3-norms are uniformly bounded, too. Therefore we may choose a subsequence of the already constructed subsequence converging also with respect to the s_2-norm and so on. In this way one obtains a sequence of sequences each of which is a subsequence of the preceding one. The Cantor diagonal sequence of this sequence of subsequences converges uniformly with respect to each B_s, $0 < s < s_0$. Denote the limit function by $\tilde{u}_* = \tilde{u}_*(t)$.

Further notice that piecewise linear functions of type (9.14) may be represented by

$$\tilde{u}(t) - \tilde{u}(t_j) = \int_{t_j}^{t} d\tau \cdot F(t_j, u(t_j))$$

provided t belongs to $[t_j, t_{j+1}]$. Analogously we have

$$\tilde{u}(t_j) - \tilde{u}(t_{j-1}) = \int_{j_{j-1}}^{t_j} d\tau \cdot F(t_{j-1}, u(t_{j-1}))$$

and so on. Adding all these equations and taking into consideration the initial condition (9.15), one obtains

$$\tilde{u}(t) - u_0 = \int_0^t d\tau \cdot F(\tau, \tilde{u}_*(\tau)) + \int_0^t d\tau \cdot \left[\frac{d\tilde{u}}{dt}(\tau) - F(\tau, \tilde{u}_*(\tau))\right] \quad (9.17)$$

because according to (9.14) the derivative of $\tilde{u}$ in $[t_j, t_{j+1}]$ is given by $F(t_j, u(t_j))$. Now replace u by the elements of the above constructed subsequence ($\tilde{u}_*$ is its limit function). Since every $\tilde{u}_k$ is piecewise linear in subsegments $[t', t'']$ of $[0, \tilde{T}]$ the integrand of the last integral may be represented in the form

$$F(t', u_k(t')) - F(\tau, \tilde{u}_*(\tau)) \quad (9.18)$$

provided τ belongs to the subsegment $[t', t'']$. Moreover take into account that the greatest length of subsegments $[t', t'']$ approaches zero, that the constructed subsequence converges uniformly, and that $F = F(t,u)$ is uniformly continuous (because a ball in $B_{s_{j+1}}$ is compact in B_{s_j}). Therefore each s-norm of the difference (9.18) approaches zero as the index approaches infinity. After replacing $\tilde{u}$ in (9.17) by the uniformly converging subsequence of the $\tilde{u}_k$ and carrying out the limiting process, we get the equation

$$\tilde{u}_*(t) - u_0 \int_0^t d\tau \cdot F(\tau, \tilde{u}_*(\tau)).$$

Hence it follwos that $\tilde{u}_*$ is a solution of the initial value problem (3.2), (3.3).

Summarizing these arguments, we see that the following theorem holds (cf. H. Begehr's paper [5] in which mainly the holomorphic case is investigated):

<u>Theorem.</u> Suppose that the imbedding operators of the scale B_s,

$0 < s < s_0$, are compact. Suppose furthermore, that the right-hand side F(t,u) of the differential equation (3.2) satisfies the conditions (I'), (II') and maps a linear function space E connected with the scale B_s into itself. Then the initial value problem (3.2), (3.3) can be solved by Euler's polygonal line method. The solution exists in $[0, \tilde{T}]$, where $\tilde{T}$ satisfies the inequality (9.16).

Concluding our considerations on Euler's polygonal method we would like to emphasize that this method yields a uniform existence interval $[0,\tilde{T}]$ for the solution, whereas the convergence of the successive approximations (3.19) can be proved only in an interval depending on s (cf. the theorem in section 3.5.)

9.7. The special case of ordinary differential equations

The initial value problem for ordinary differential equations (cf. section 1.5.) may be interpreted as special case of the theorem on the solution of initial value problems in scales of Banach spaces. For this end we have only to replace the scale B_s, $0 < s < s_0$, of Banach spaces by a fixed Banach space B. Therefore the abstract theorems of chapter 3 (see the sections 3.5. and 3.8.) may be regarded as uniform generalization of theorems on initial value problems for both ordinary and partial differential equations.

9.8. Initial value problems for equations with singular coefficients

A completely new situation arises from a zero of a factor of the derivative with respect to the time. Solving such an equation for the derivative, one obtains an equation whose coefficients possess singularities. If the factor of the derivative to t vanishes at t = 0, then the initial value problem is not solvable, in general. In some cases (see [63]) the initial condition can be replaced by a boundary condition. An abstract non-linear Cauchy-Kovalevskaya theorem for equations with singular coefficients was formulated and proved by M. Reissig in his paper [52]. It may be added that the same author proves existence and uniqueness theorems for equations with coefficients which are powers of t by using partial Fourier transformations in [53].

An algebraic characterization of the solvability in the linear holomorphic case with singular coefficients was given by T. Oshima [49]. Denote the independent variables (inclusivily t) by $z_1, \ldots, z_n$, $z = (z_1, \ldots, z_n)$. Regard the equation

$$\sum_{i=1}^{n} a_i(z)\frac{\partial u}{\partial z_i} + b(z)u = f(z), \tag{9.19}$$

where the coefficients are holomorphic in $z_1, \ldots, z_n$. The classical Cauchy-Kovalevskaya theorem is applicable provided the ideal $\mathfrak{a}$ generated by $a_1(z), \ldots, a_n(z)$ is equal to the whole ring of holomorphic functions. T. Oshima investigates the case that $\mathfrak{a}$ is a proper ideal. The equation (9.19) is solvable if and only if f belongs to the ideal $\mathfrak{b}$ generated by $a_1(z), \ldots, a_n(z), b(z)$. The solution is uniquely determined if $\mathfrak{a} \neq \mathfrak{b}$. In the case $\mathfrak{a} = \mathfrak{b}$ there is a one-one correspondence between the solutions and the Cauchy data on the variety defined by $\mathfrak{b}$.

In order to show that the uniqueness of the solution of initial value problems can get lost in the case of singular coefficients, we regard the differential equation

$$z_1\frac{\partial u}{\partial z_1} - z_2\frac{\partial u}{\partial z_2} = 0.$$

Then

$$u_1 = u_1(z) = 0$$

and

$$u_2 = u_2(z) = \sin(z_1 \cdot z_2)$$

are two different solutions of the initial value problem

$$u(0,z_2) = 0.$$

9.9. Cauchy-Kovalevskaya theorems for a vector-valued time variable

In all preceding considerations t was a real (or complex) variable. Recently a Cauchy-Kovalevskaya theorem for a vector-valued time variable t has been proved. Before explaining this result and sketching its proof, we shall point out some features of the case of a vector-valued time variable.

First regard the case of m desired functions $u_k = u_k(t_1,\ldots,t_n)$, $k = 1, \ldots, m$, depending on n real variables $t_1, \ldots, t_n$ and satisfying the system

$$\frac{\partial u_k}{\partial t_i} = f_{ki}(t,u), \tag{9.20}$$

where $t = (t_1, \ldots, t_n)$, $u = (u_1, \ldots, u_m)$, $i = 1, \ldots, n$. Let $u = u(t)$ be a twice continuously differentiable solution. Differentiating (9.20) with respect to t_j and using the chain rule, one obtains

$$\frac{\partial^2 u_k}{\partial t_i \partial t_j} = \frac{\partial f_{ki}}{\partial t_j} + \sum_{\mu=1}^{m} \frac{\partial f_{ki}}{\partial u_\mu}\frac{\partial u_\mu}{\partial t_j} = \frac{\partial f_{ki}}{\partial t_j} + \sum_{\mu=1}^{m} \frac{\partial f_{ki}}{\partial u_\mu} f_{\mu j}. \tag{9.21}$$

On the other hand, the second order derivatives do not depend on the order in which the differentiations are carried out, i.e.,

$$\frac{\partial^2 u_k}{\partial t_i \partial t_j} = \frac{\partial^2 u_k}{\partial t_j \partial t_i} .$$

Comparing the right-hand sides of (9.21) and of the corresponding equation for

$$\frac{\partial^2 u_k}{\partial t_j \partial t_j} ,$$

it follows

$$\frac{\partial f_{ki}}{\partial t_j} + \sum_{\mu=1}^{m} \frac{\partial f_{ki}}{\partial u_\mu} f_{\mu j} = \frac{\partial f_{kj}}{\partial t_i} + \sum_{\mu=1}^{m} \frac{\partial f_{kj}}{\partial u_\mu} f_{\mu i}. \tag{9.22}$$

Let $u = u(t)$ be a solution to (9.20) satisfying the initial condition

$$u(t_0) = c, \tag{9.23}$$

where

$$t_0 = (t_{o1}, \ldots, t_{on})$$

and

$$c = (c_1, \ldots, c_m).$$

Define

$$\tau = \frac{1}{\lambda}(t - t_0)$$

and

$$v(\lambda,\tau) = u(t_0 + \lambda\tau).$$

Interpreting $\tau = (\tau_1, \ldots, \tau_n)$ as parameter, the vector $v = (v_1,\ldots,v_m)$ turns out to be a solution to the system of ordinary differential equations

$$\frac{dv_k}{d\lambda} = \sum_{\nu=1}^{n} \frac{\partial u_k}{\partial t_\nu} \tau_\nu,$$

i.e.,

$$\frac{dv_k}{d\lambda} = \sum_{\nu=1}^{n} \tau_\nu f_{k\nu}(t_0 + \lambda\tau), \tag{9.24}$$

satisfying the initial condition

$$v(0,\tau) = u(t_0) = c \tag{9.25}$$

for each τ.

Conversely, let $v = v(\lambda,\tau)$ be the solution of the initial value problem (9.24), (9.25), where τ is a fixed parameter. For $\tau = (0, \ldots, 0)$ the system (9.24) passes into

$$\frac{dv_k}{d\lambda} = 0,$$

i.e., one has

$$v(\lambda,0) = v(0,0) = c \tag{9.26}$$

for each λ.

Now define

$$u(t) = v(1, t-t_0). \tag{9.27}$$

In view of (9.26) it follows

$$u(t_0) = v(1,0) = c,$$

i.e., $u = u(t)$ satisfies the initial condition (9.23). A. Mayer proved in his paper [37] that the vector $u = u(t)$ defined by (9.27) satisfies the first order system (9.20), too, provided its right-hand sides f_{ki} satisfy the relations (9.21) (cf. also E. Goursat [27] and C. Carathéodory [11]). Systems of type (9.20) satisfying the relations (9.22) are called <u>completely integrable</u>. A. Mayer's construction shows that for completely integrable systems of type (9.20) the initial value problem (9.20), (9.23) can be reduced to the initial value problem (9.25) for the system (9.24) of ordinary differential equations depending on the parameter τ.

Let us remark that analogous constructions can be carried out also for completely integrable systems in several complex variables (see [62]).

Second take into consideration that the differential calculus can be generalized to the case of variables belonging to Banach spaces (see J. Dieudonné [20] and S. Lang [32]). Then a derivative is interpreted as linear mapping between the Banach spaces under consideration, whereas higher order derivatives can be interpreted as multilinear mappings. The independence of higher order derivatives of the order of the differentiations amounts to symmetry properties of the corresponding multilinear mappings.

If the variables t and u belong to Banach spaces, the system (9.20) is to be replaced by the more general differential equation

$$\frac{du}{dt} = f(t,u). \tag{9.28}$$

Then the integrability condition of (9.28) can be written as symmetry relation for the partial derivatives of the right-hand side $f(t,u)$. Then the so-called <u>Frobenius theorem</u> states that the initial value problem

$$u(t_0) = u_0 \tag{9.29}$$

for the system (9.28) is uniquely solvable provided the system under consideration is completely integrable. Again substituting $t = t_0 + \lambda\tau$, the initial value problem (9.28), (9.29) can be reduced to the initial value problem

$$v(0,\tau) = u_o$$

for the Banach-space-valued function

$$v = v(\lambda,\tau) = u(t_o + \lambda\tau)$$

satisfying an ordinary first order differential equation in the (real) variable τ. The desired solution $u = u(t)$ of (9.28), (9.29) is obtained by setting

$$u(t) = v(1, t-t_o).$$

The differential equation (9.28) can be interpreted as generalization of the differential equation (0.1) to the case that the values $u(t)$ of the desired solution $u = u(t)$ as well as the independent variable t belong to Banach spaces. The Cauchy-Kovalevskaya theorem deals with the more general differential equation (0.3) whose right-hand side depends also on the derivative with respect to a further variable x. Such generalization is also possible for variables running in Banach spaces. Strictly speaking, we look for a solution $u = u(t,x)$ of the initial value problem

$$D_t u = f(t,x,u,D_x u), \tag{9.30}$$

$$u(t_o,x) = u_o(x), \tag{9.31}$$

where the variables t, x, u belong to given Banach spaces and $D_t u$ and $D_x u$ denote the partial derivatives to the variables t and x resp.

Suppose the differential equation (9.30) having a smooth right-hand side is completely integrable, i.e., the right-hand side satisfies a certain symmetry condition for the first order derivatives generalizing the analogous one in the case of the simpler differential equation (9.28). Suppose further that the variables x and u belong to complex (not real) Banach spaces (this implies that the right-hand side is analytic in those variables);

Then the following generalization of the Cauchy-Kovalevskaya theorem to the case of a vector-valued time variable t holds:

> The initial value problem (9.30), (9.31) possesses a unique Banach-space valued solution.

This theorem has been proved by T. Yamanaka by reducing it to a differential equation with a scalar-valued time variable and a vector-valued parameter. Detailed proofs can be found in T. Yamanaka's paper [81]. There are also given some comments on similar results of M. Shinbrot and R. R. Welland [59] and of L. B. Mogilevskaya [41, 42]. T. Yamanaka points out, too, that his result includes B. Lascar's Cauchy-Kovalevskaya theorem [33] for an infinite dimensional space variable and a scalar valued time variable. Let us additionally refer to L. B. Mogilevskaya's new paper [43].

10. FURTHER UNIQUENESS THEOREMS

10.1. Uniqueness theorems for initial value problems in higher dimensions

In a similar way as done in section 7.3. for the case of the plane, also in higher dimensions uniqueness theorems can be immediately deduced from the Holmgren theorem and its generalization. Regard, for instance, two solutions u, $\tilde{u}$ to the initial value problem (9.12). Then the difference $u - \tilde{u}$ is identically equal to zero at $t = 0$ and satisfies the differential equation

$$\frac{\partial(u - \tilde{u})}{\partial t} = L_0(u - \tilde{u}),$$

where L_0 is the homogeneous part of the differential operator (9.11),

$$L_0 u = \left(\sum_{i,j} A^1_{ij} \frac{\partial u_i}{\partial x_j} + \sum_i B^1_i u_i, \ldots\right).$$

Applying the classical Holmgren theorem (see 5.1.) and its generalization (see 5.4.), the following uniqueness theorem follows immediately:

Suppose the coefficients A^K_{ij}, B^K_i of the differential operator (9.11) are power series in t, x_1, x_2, x_3 or at least deformed power series. Then the initial value problem (9.12) possesses at most one solution.

10.2. Permanence principles

A further question connected with initial value problems of type (9.12) is the following one:

Does the solution $u = u(t,x)$ satisfy any side condition $lu(t,\cdot) = 0$ at each t provided the initial function u_0 does so?

This question leads to the concept of permanence principles:

A differential operator l whose coefficients are independent of t is said to be a permanence principle to the differential equation

$$\frac{\partial u}{\partial t} = Lu \tag{10.1}$$

if the side condition

$$lu(t,\cdot) = 0$$

is satisfied at each t provided it is satisfied by the initial function u_0.

Now we are goint to explain two kinds of permenence principles for equations of type (10.1) (cf. [71]). In both cases we assume that a uniqueness theorem holds for the initial value problem (9.12).

a) First assume that L and l are commutative. Then $lu = 0$ is a permanence principle to the differential equation (10.1).

Proof: Denoting lu by U, we obtain

$$\frac{\partial U}{\partial t} = \frac{\partial}{\partial t}(lu) = l\left(\frac{\partial u}{\partial t}\right) = l(Lu) = L(lu) = LU.$$

Therefore the assumption $U(0,\cdot) = 0$ implies $U(t,\cdot) = 0$ at each t.

b) Second assume that L is a generalized Cauchy-Riemann operator (see 2.3.) associated to l (notice that commutative operators are associated). Then the initial value problem (9.12) is solvable in a scale defined by the side condition $lu = 0$. Let u be this solution, constructed by the method of scales of Banach spaces. Then the side condition $l\tilde{u}(t,\cdot) = 0$ is satisfied at each t. Now let $u = u(t,x)$ be any solution to the same initial value problem (9.12). Then the difference $u - \tilde{u}$ is solution to the homogeneous equation

$$\frac{\partial(u - \tilde{u})}{\partial t} = L_o(u - \tilde{u}) \tag{10.2}$$

vanishing at $t = 0$. Applying the uniqueness theorem, it follows that u and $\tilde{u}$ coincide at each t. Thus the side condition $lu(t,\cdot) = 0$ is satisfied at each t, i.e., l is a permanence principle.

E.g., the differential equations (9.10) for potential vectors describe a permanence principle for the differential equation (10.1) provided L is associated to the differential operator which is defined by the left-hand sides of (9.10) (cf. 9.4.1.). In the paper [72], moreover, constants λ_1, λ_2, λ_3, μ_1, μ_2, μ_3 are determined for which the differential operator rot is associated to the differential equations

$$\operatorname{div} u = 0, \qquad \sum_{i=1}^{3} \lambda_i \frac{\partial u_i}{\partial x_i} = 0,$$

$$\frac{\partial u_2}{\partial x_1} = \mu_3 \frac{\partial u_1}{\partial x_2}, \qquad \frac{\partial u_3}{\partial x_2} = \mu_1 \frac{\partial u_2}{\partial x_3}, \qquad \frac{\partial u_1}{\partial x_3} = \mu_2 \frac{\partial u_3}{\partial x_1}.$$

Note, finally, that in the case b) we need the uniqueness theorem only for the homogeneous equation (10.2) but not for the possibly inhomogeneous equation (10.1). Permanence principles hold also in the case of more than three spacelike variables.

10.3. Uniqueness in dependence on the scale

A uniqueness theorem for an initial value problem in a scale of Banach spaces states that there exists at most one solution. This does not exclude that there exist further solutions of the same initial value problem not belonging to the scale under consideration.

M. Reissig [54] constructs a scale in which a suitably posed initial value problem is uniquely solvable, whereas the uniqueness get lost in an extended scale. Therefore, an absolute uniqueness theorem cannot be expected. In some cases, however, the uniqueness is preserved in an extended class of functions.

The classical Holmgren theorem (see 5.1.) shows, for instance, that in the holomorphic case other solutions than the holomorphic ones do not exist. Moreover, in A. Crodel's thesis [16] it is proved that the uniqueness theorem remains true in an extended scale defined by a differential inequality.

10.4. A generalized Gronwall lemma for differential inequalities in scales of Banach spaces

A. Crodel shows in his thesis [16] (see also [19]) that uniqueness theorems for initial value problems in scales of Banach spaces can also be proved by using the following generalized Gronwall lemma:

Theorem. Suppose that $u = u(t)$ is a continuously differentiable mapping of $[0, T)$ into a scale B_s, $0 < s < s_0$. Suppose further that for s, s' with $0 < s' < s < s_0$ the inequality

$$\left\| \frac{du}{dt}(t) \right\|_{s'} \leqq \frac{C}{s - s'} \| u(t) \|_s \tag{10.3}$$

holds, where C does not depend on s and s'. Then

$$u(t) = 0$$

in B_s for each t with

$$0 \leqq t < \min \left(T, \frac{s_0 - s}{Cs} \right).$$

Sketch of the proof. Take any point with $u(t_0) = 0$ in B_s. Then

$$u(t) = \int_{t_0}^{t} d\tau \cdot \frac{du}{dt}(\tau)$$

and, consequently,

$$\| u(t) \|_{s'} \leqq \int_{t_0}^{t} \left\| \frac{du}{dt}(\tau) \right\|_{s'} d\tau . \tag{10.4}$$

Applying the inequality (10.3), one obtains

$$\| u(t) \|_{s'} \leqq \frac{C|t - t_0|}{s - s'} \sup_{\tau \in [t_0, t]} \| u(\tau) \|_s$$

$$\leqq \frac{Ce|t - t_0|}{s - s'} \sup_{\tau \in [t_0, t]} \| u(\tau) \|_s .$$

Therefore, the following inequality is true for $k = 1$:

$$\|u(t)\|_{s'} \leq \left(\frac{Ce|t - t_0|}{s - s'}\right)^k \sup_{\tau \in [t_0, t]} \|u(t)\|_s. \tag{10.5}$$

In order to prove it for an arbitrary $k = 1, 2, \ldots$, take any pair s, s' with $0 < s' < s < s_0$ and define

$$\tilde{s} = s' + \frac{s - s'}{k + 1}.$$

Then (10.5) implies

$$\|u(\tau)\|_{\tilde{s}} \leq \left(\frac{Ce|\tau - t_0|}{s - \tilde{s}}\right)^k \sup_{\tau' \in [t_0, \tau]} \|u(\tau')\|_s$$

$$\leq \left(\frac{Ce|\tau - t_0|}{s - s'}\right)^k \cdot e \cdot \sup_{\tau' \in [t_0, \tau]} \|u(\tau')\|_s \tag{10.6}$$

because

$$\left(\frac{k + 1}{k}\right)^k = \left(1 + \frac{1}{k}\right)^k < e.$$

Once more taking into consideration the assumption (10.3) with s' and $\tilde{s}$, the estimate (10.4) leads to

$$\|u(t)\|_{s'} \leq \frac{C}{\tilde{s} - s'} \int_{t_0}^{t} \|u(\tau)\|_{\tilde{s}} d\tau.$$

In view of (10.6) we obtain, finally,

$$\|u(t)\|_{s'} \leq \left(\frac{Ce|t - t_0|}{s - s'}\right)^{k+1} \sup_{\tau' \in [t_0, t]} \|u(\tau')\|_s,$$

i.e., (10.5) holds for $k + 1$, too.

Letting k tend to infinity in (10.5), we see that $u(t) = 0$ in a neighbourhood of t_0. This proves the above theorem.

A. Crodel's considerations yield a new (and simpler) proof of the theorem in section 3.7. This uniqueness theorem turns out to be a special case of the above formulated generalized Gronwall lemma because for two solutions $u = u(t)$ and $v = v(t)$ of (3.2), (3.3) one gets the equation

$$\frac{d}{dt}(u(t) - v(t)) = F(t, u(t)) - F(t, v(t))$$

and, therefore,

$$\left\|\frac{d}{dt}(u(t) - v(t))\right\|_{s'} \leq \frac{C}{s - s'}\|u(t) - v(t)\|_s.$$

REFERENCES

[1] A. A. Agrachev and S. A. Vakhramev (А. А. Аграчев и С. А. Вахрамеев), Хронологические ряды и теорема Коши-Ковалевской. Итоги науки и техн. ВИНИТИ. Проблемы геометрии, т. 12, 165 - 189, 1981.

[2] P. S. Aleksandrov (П. С. Александров), Введение в теорию множеств и общую топологию. Москва 1977.

[3] P. S. Aleksandrov and O. A. Oleinik (eds. П. С. Александров и О. А. Олейник), Труды всесоюзной конференции по уравнениям с частными производными. Изд. Московского унив., 1978.

[4] M. S. Baouendi and C. Goulaonic, Sharp estimates for analytic pseudodifferential operators and applications to Cauchy problems. Journ. Diff. Equat., vol. 48, № 2, 241 - 268, 1983.

[5] H. Begehr, Eine Bemerkung zum nichtlinearen klassischen Satz von Cauchy-Kowalewski. Math. Nachr., Bd. 131, 175 - 181, 1987.

[6] -, Der Satz von Cauchy-Kowalewski für hyperanalytische Funktionen. Zeitschr. Anal. Anwend., Bd. 6, Heft 1, 43 - 47, 1987.

[7] L. Bers, Theory of Pseudoanalytic Functions. New York University 1953.

[8] L. Bers, F. John, and M. Schechter, Partial differential equations. New York / London / Sidney 1964. Russian transl.: Л. Берс, Ф. Джон и М. Шехтер, Уравнения с частными производыми. Москва 1966.

[9] B. V. Bojarski, Теория обобщенного аналитического вектора. Ann. Polon. Math., vol. 17, 281 - 320, 1966.

[10] F. Brackx, R. Delanghe, and F. Sommen, Clifford Analysis. Boston / London / Melbourne 1982.

[11] C. Carathéodory, Variationsrechnung und partielle Differentialgleichungen erster Ordnung. Bd. I., 2. Aufl., Leipzig 1956.

[12] D. L. Colton, Analytic Theory of Partial Differential Equations. Boston / London / Melbourne 1980.

[13] R. Courant and D. Hilbert, Methods of Mathematical Physics, Vol. II. New York / London 1962. Russian transl.: Р. Курант, Уравнения с частными производными. Москва 1964.

[14] A. Crodel, Nichtlineare Evolutionsgleichungen für q-holomorphe Vektoren. Zeitschr. Anal. Anwend., Bd. 6, Heft 1, 49 - 59, 1987.

[15] -, A-priori estimates for generalized q-holomorphic vectors.

[16] -, Sätze vom Cauchy-Kowalewskaja Typ für partielle komplexe Differentialgleichungssysteme in Klassen verallgemeinerter analyti-

scher Vektoren. Thesis (Dissertation A), Universität Halle, 1986.

[17] -, Sätze vom Cauchy-Kowalewskaja Typ für lineare Anfangswertprobleme in Klassen verallgemeinerter analytischer Vektoren (to appear).

[18] -, Sätze vom Cauchy-Kowalewskaja Typ für nichtlineare Anfangswertprobleme in Klassen verallgemeinerter analytischer Vektoren (to appear).

[19] -, Ein verallgemeinertes Gronwallsches Lemma in Banachraumskalen (to appear).

[20] J. Dieudonné, Foundations of Modern Analysis, 7th ed. New York / London 1968. German transl.: Grundzüge der modernen Analysis, Berlin 1971. Russ. transl.: Основы современного анализа. Москва 1964.

[21] J. Duchon et R. Robert, Estimation d'operateurs integraux du type de Cauchy dans les echelles d'Ovsjannikov et application. C. R. Acad. Sc. Paris, t. 299, Série I, № 13, 595 - 598, 1983.

[22] V. R. Friedlender (В. Р. Фридлендер), Об аналитических решениях задачи Коши для некоторых нелинейных уравнений с частными производными. Мат. Сборн., т. 47(89), № 1, 17 - 44, 1959.

[23] A. Friedman, A new proof and generalizations of the Cauchy-Kowalewski theorem. Transact. Amer. Math. Soc., vol. 98, 1 - 20, 1961.

[24] I. M. Gelfand and G. E. Shilov (И. М. Гельфанд и Г. Е. Шилов), Обобщенные функции, т. 111, Москва 1958. German transl.: Verallgemeinerte Funktionen, Bd. III, Berlin 1964.

[25] R. P. Gilbert, Constructive Methods for Elliptic Equations. Lecture Notes in Mathematics, vol. 365. Berlin / Heidelberg / New York 1974.

[26] B. Goldschmidt, Funktionentheoretische Eigenschaften verallgemeinerter analytischer Vektoren. Math. Nachr., Bd. 90, 57 - 90, 1979.

[27] E. Goursat, Leçons sur l'intégration des équations aux derivées partielles du premier ordre. Deuxième édition Paris 1921.

[28] L. Hörmander, Linear Partial Differential Operators. Berlin / Göttingen / Heidelberg 1963. Russ. transl.: Линейные дифференциальные операторы с частными производными. Москва 1965.

[29] F. John, Partial differential equations, 4th ed. New York / Heidelberg / Berlin 1982.

[30] K. Keller and A. Schneider, Ein funktionalanalytischer Beweis des Satzes von Cauchy-Kowalewsky. Manuscripta Math., vol. 39, № 1, 31 - 37, 1982.

[31] E. Lanckau and W. Tutschke (eds.), Complex Analysis. Methods, Trends, and Applications, Berlin 1983.

[32] S. Lang, Real Analysis. Reading 1969.

[33] B. Lascar, Théorème de Cauchy-Kovalewsky et théorème d'unicité d'Holmgren pour fonctions d'une infinité de variables. C. R. Acad. Sc. Paris, t. 282, 691 - 694, 1976.

[34] J. Le Roux, Sur les intégrales analytiques de l'équation $\partial^2 u/\partial y^2 = \partial u/\partial x$. Bull. Sci. Math. France, vol. 19, 127 - 129, 1895.

[35] H. Lewy, An example of a smooth linear partial differential equation without solution. Ann. of Math., vol. 66, 155 - 158, 1957.

[36] G. F. Mandzhavidse (Г. Ф. Манджавидзе) and W. Tutschke, Некоторые оценки норм производных голоморфных функций и их применение к задачам с начальными значениями. Zeitschr. Anal. Anwend., Bd. 3, 1 - 5, 1984.

[37] A. Mayer, Über unbeschränkt integrable Systeme von linearen totalen Differentialgleichungen. Math. Ann., Bd. 5, 448 - 470, 1872.

[38] E. J. McShane, Extension of range functions. Bull. Amer. Math. Soc., vol. 40, 837 - 846, 1934.

[39] S. Mizohata, Une remarque sur le théorème de Cauchy-Kowalewski. Ann. Scuola Norm. Sup. Pisa Cl. Sci. (4) 5, № 3, 559 - 566, 1978.

[40] -, On the Cauchy Kowalewski theorem. Math. analysis and applications, Part B, 617 - 652. Adv. in Math. Suppl. Stud., 7 b, Academic Press, New York, 1981.

[41] L. B. Mogilevskaya (Л. Б. Могилевская), Об аналитичности решений дифференциальных уравнений в Банаховом пространстве. Доклады АН СССР, т. 228, № 1, 30 - 33, 1976. English transl.: On the analyticity of solutions of differential equations in Banach space. Soviet Math. Dokl., vol. 17, 643 - 646, 1976.

[42] -, Correction to the above paper. Доклады АН СССР, т. 231, № 3, 520, 1976. Engl. transl.: Soviet Math. Dokl., vol 17, № 4, p. V, 1977.

[43] -, Теорема Коши-Ковалевской для Банахова пространства. Теория операторов в функциональных пространствах, 48 - 64. Издат. Воронежского универс. Воронеж 1983.

[44] M. Nagumo, Über das Anfangswertproblem partieller Differentialgleichungen. Japan. Journ. Math., vol. 18, 41 - 47, 1941.

[45] L. Nirenberg, An abstract form of the nonlinear Cauchy-Kowalewski theorem. Journ. Diff. Geom., vol. 6, 561 - 576, 1972.

[46] L. Nirenberg, Topics in Nonlinear Functional Analysis. New York 1974. Russian transl.: Лекции по нелинейному функциональному анализу. Москва 1977.

[47] T. Nishida, A Note on Nirenberg's Theorem as an Abstract Form of the Nonlinear Cauchy-Kowalewski Theorem in a Scale of Banach Spaces. Journ. Diff. Geom., vol. 12, 629 - 633, 1977.

[48] H. Okamura, On the Cauchy-Kowalewski-Nagumo Theorem (in Japanese). Functional Equations and Applied Analysis, № 29, 20 - 34, 1941.

[49] T. Oshima, On the theorem of Cauchy Kowalewsky for first order linear differential equation with degenerate principal symbol. Proc. Jap. Acad., vol. 49, 83 - 87, 1973.

[50] L. V. Ovsyannikov (Л. В. Овсянников), Сингулярный оператор в шкале банаховых пространств. Доклады АН СССР, т. 163, № 4, 819 - 822, 1965. English transl.: Singular Operator in a Scale of Banach Spaces. Soviet Math. Dokl., vol. 6, 1025 - 1028, 1965

[51] -, Задача Коши в шкале банаховых пространств аналитических функций. Труды симпозиума по механике сплошной среды и родственным проблемам анализа (Тбилиси 23. - 29. 9. 1971), т. 11, 219 - 229. Тбилиси 1974.

[52] M. Reissig, Ein abstraktes nichtlineares Cauchy-Kowalewskaja Theorem mit singulären Koeffizienten. I: Zeitschr. Anal. Anwend., Bd. 6, Heft 1, 35 - 41, 1987. II: Zeitschr. Anal. Anwend., Bd. 7, Heft 2, 171 - 183, 1988.

[53] -, Anwendung der Methode der partiellen Fouriertransformation auf die Lösung partieller Differentialgleichungen mit stark singulären Koeffizienten (to appear).

[54] -, About the theorem of Holmgren for abstract Cauchy-Kovalevsky problems (to appear).

[55] W. Rüprich, Das Rand-Sprungwertproblem für implizite partielle komplexe Differentialgleichungssysteme in der Ebene. Math. Nachr., Bd. 101, 309 - 319 (1981).

[56] -, Eine Lösungsmethode für das Randwert-Sprung-Problem für implizite partielle Differentialgleichungssysteme in der Ebene. Math. Nachr., Bd. 118, 167 - 178 (1984).

[57] -, Funktionalanalytische Methoden bei räumlichen und zeitabhängigen impliziten Differentialgleichungsproblemen. Thesis (Dissertation A), Universität Halle, 1985.

[58] T. Sekine and T. Yamanaka, On a functional differential equation in

a scale of Banach spaces. Report of the Research Institute of Science and Technology, Nihon University, № 24, 1 - 8, 1980.

[59] M. Shinbrot and R. R. Welland, The Cauchy-Kowalewskaya Theorem. Journ. Math. Anal. Applic., vol. 55, 757 - 772, 1976.

[60] A. N. Tikhonov (А. Н. Тихонов), Über unendliche Systeme von Differentialgleichungen. Матем. Сборник, т. 41, 551 - 560 (1934).

[61] F. Treves, Basic Linear Partial Differential Equations. New York / San Francisco / London 1975.

[62] W. Tutschke, Partielle komplexe Differentialgleichungen in einer und in mehreren komplexen Variablen. Berlin 1977.

[63] -, Elliptic Evolution Processes. Math. Nachr., Bd. 100, 21 - 31 (1981).

[64] -, Partielle Differentialgleichungen. Klassische, funktionalanalytische und komplexe Methoden. Teubner-Text zur Mathematik, Bd. 27, Leipzig 1983.

[65] -, On an abstract nonlinear Cauchy-Kowalewski theorem - a variant of L. Nirenberg's and T. Nishida's proof. Zeitschr. Anal. Anwend., Bd. 5, Heft 2, 185 - 192, 1986.

[66] -, Задача с начальными значениями для обобщенных аналитических функций, зависящих от времени (Обобщения теорем Коши-Ковалевской и Хольмгрена). Доклады АН СССР, т. 262, № 5, 1081 - 1085, 1982. English transl.: V. Tučke, A problem with Initial Values for Generalized Analytic Functions Depending on Time (Generalizations of the Cauchy-Kowalewski and Holmgren Theorems). Soviet Math. Dokl., vol. 25, № 1, 201 - 205, 1982.

[67] -, Ассоциированныи операторы комплексного анализа. Сообщ. АН ГрузССР, т. 107, № 3, 481 - 484, 1982.

[68] -, Потенциальные векторы, зависящие от времени. Дифференциальные уравнения в частных производных и их приложения. Труды всесоюзного симпозиума в Тбилиси 21 - 23 апреля 1982 г., 235 - 239. Тбилиси 1986.

[69] -, Решение задачи Коши в классах функций, являющихся пространственными обобщениями обобщенных аналитических функций. Матем. структуры - вычислительная математика - матем. моделирование, т. 2, 85 - 89. София 1984.

[70] -, Cauchy Problems with Monogenic Initial Values. In: Analytic Functions, Błażejewko 1982, Proceedings, ed. by J. Ławrynowicz. Lecture Notes in Mathematics, vol. 1039, 453 - 457. Berlin / Heidelberg / New York / Tokyo 1983.

[71] W. Tutschke, Permanence Principles for First Order Differential Operators. Bull. Soc. Scienc. Lettr. Łódź, vol. XXXIV, № 10, 1984.

[72] -, First Order Differential Operators in the Threedimensional Euclidian Space Being Associated to the Differential Operator rot. Complex Variables: Theory and Application; vol. 7, 349 - 355, 1987.

[73] -, An abstract non-linear Cauchy-Kowalewski theorem and its proof by a contraction-mapping principle. Математички весник 38, 597 - 607, 1986.

[74] -, Construction of generalized analytic functions depending on time by a contraction-mapping principle. Proceed. Conf. on Modern Problems in Math. Physics, Tbilissi 1987 (in print).

[75] W. Tutschke and C. Withalm, The Cauchy-Kovalevska Theorem for Pseudoholomorphic Functions in the Sense of L. Bers. Complex Variables: Theory and Applications; vol. 1, 389 - 393, 1983.

[76] K. G. Valeev and O. A. Zhautykov (К. Г. Валеев и О. А. Жаутыков), Бесконечные системы дифферинциальных уравнений. Алма Ата 1974.

[77] I. N. Vekua (И. Н. Векуа), Обобщенные аналитические функции. Москва 1959. English transl.: Generalized Analytic Functions. Oxford / Reading 1962. Translation into German: Verallgemeinerte analytische Funktionen. Berlin 1963.

[78] W. Walter, An elementary proof of the Cauchy-Kowalevsky theorem. Amer. Math. Monthly, vol. 92, 115 - 125, 1985.

[79] -, Functional differential equations of the Cauchy-Kowalevsky type. Aequat. Mathem., vol. 28, 102 - 113, 1985.

[80] T. Yamanaka, Note on Kowalewskaja's System of Partial Diff. Equations. Comment. Math. Univ. St. Pauli, vol. 9, 7 - 10, 1961.

[81] -, The Cauchy-Kovalevskaja Theorem with a vector valued time variable. Funkcialaj Ekvacioj (Serio Internacia), vol. 24, numero 2, 211 - 246, 1981.

[82] T. Yamanaka and H. Tamaki, Cauchy-Kovalevskaja Theorem for functional partial differential equations. Comment. Math. Univ. St. Pauli, vol. 29, № 1, 55 - 64, 1980.

INDEX

MIX
Papier aus verantwortungsvollen Quellen
Paper from responsible sources
FSC® C105338

If you have any concerns about our products,
you can contact us on
ProductSafety@springernature.com

In case Publisher is established outside the EU,
the EU authorized representative is:
Springer Nature Customer Service Center GmbH
Europaplatz 3, 69115 Heidelberg, Germany

Printed by Libri Plureos GmbH
in Hamburg, Germany